Multirate and Multiphase
Switched-capacitor Circuits

ML

UNIVERSITY OF STRATHCLYDE

30125 00543695 0

**Books are to be returned on or before
the last date below.**

Multirate and Multiphase Switched-capacitor Circuits

Adam Dąbrowski
Institute of Electronics and Telecommunication
Poznań University of Technology
Poznań, Poland

CHAPMAN & HALL
London · Weinheim · New York · Tokyo · Melbourne · Madras

Published by Chapman & Hall, 2–6 Boundary Row, London SE1 8HN, UK

Chapman & Hall, 2–6 Boundary Row, London SE1 8HN, UK

Chapman & Hall GmbH, Pappelallee 3, 69469 Weinheim, Germany

Chapman & Hall USA, 115 Fifth Avenue, New York, NY 10003, USA

Chapman & Hall Japan, ITP-Japan, Kyowa Building, 3F, 2-2-1 Hirakawacho, Chiyoda-ku, Tokyo 102, Japan

Chapman & Hall Australia, 102 Dodds Street, South Melbourne, Victoria 3205, Australia

Chapman & Hall India, R. Seshadri, 32 Second Main Road, CIT East, Madras 600 035, India

First edition 1997

Published by arrangement with Polish Scientific Publishers PWN Ltd.

© 1997 Polish Scientific Publishers PWN, Warszawa

Printed in Great Britain at T.J. Press (Padstow) Ltd., Padstow, Cornwall

ISBN 0 412 72490 1

Apart from any fair dealing for the purposes of research or private study, or criticism or review, as permitted under the UK Copyright Designs and Patents Act, 1988, this publication may not be reproduced, stored, or transmitted, in any form or by any means, without the prior permission in writing of the publishers, or in the case of reprographic reproduction only in accordance with the terms of the licences issued by the Copyright Licensing Agency in the UK, or in accordance with the terms of licences issued by the appropriate Reproduction Rights Organization outside the UK. Enquiries concerning reproduction outside the terms stated here should be sent to the publishers at the London address printed on this page.

The publisher makes no representation, express or implied, with regard to the accuracy of the information contained in this book and cannot accept any legal responsibility or liability for any errors or omissions that may be made.

A Catalogue record for this book is available from the British Library

∞ Printed on permanent acid-free text paper, manufactured in accordance with ANSI/NISO Z39.48-1992 and ANSI/NISO Z39.48-1984 (Permanence of Paper).

To my parents *Anna* and *Mirosław*

To my wife *Agata*
and my son *Marcin*

Contents

			Preface	xi
1	**Introduction**			**1**
	1.1	Modern versus classical filters		1
	1.2	Switched-capacitor filters		5
	1.3	Multirate systems		12
2	**Basic elements of SC circuits**			**15**
	2.1	Switches and clock		16
	2.2	Capacitors		22
	2.3	Amplifiers		27
3	**Discrete-time systems**			**31**
	3.1	Discrete-time signals		31
		3.1.1	Sampling process	31
		3.1.2	Quantization	33
		3.1.3	Fourier representation of discrete-time signals	35
		3.1.4	Sampling theorems	38
		3.1.5	Modified $\mathcal{Z}$ and Dirichlet transformation	42
		3.1.6	Polyphase signal decomposition	46
	3.2	Discrete-time pseudopassive systems		49
		3.2.1	Wave discrete-time systems	49
		3.2.2	Discrete-time reference circuit	51
		3.2.3	Pseudopower and pseudoenergy	52
		3.2.4	Pseudopassivity and pseudolosslessness	55
4	**Multirate systems**			**61**
	4.1	Sampling rate alteration		61
		4.1.1	Interpolation	62
		4.1.2	Decimation	65
	4.2	Product modulation		67

viii *Contents*

	4.3	Multirate system concept	73
	4.4	Analysis of multirate systems	75
		4.4.1 Systems with periodically varying parameters	75
		4.4.2 Construction of frequency characteristics	79
		4.4.3 Equivalences for building-block connections	82
		4.4.4 Polyphase representation	85
	4.5	Filter banks	91
		4.5.1 Branching filter banks	93
		4.5.2 Quadrature-mirror filter banks	98
		4.5.3 Multichannel filter banks	108
	4.6	Multirate wave discrete-time arrangements	111
		4.6.1 Multirate lattice arrangements	112
		4.6.2 Multirate ladder arrangements	118
5	**Systems with nonuniformly sampled signals**		**119**
	5.1	Nonuniform sampling analysis	120
		5.1.1 Commensurate sampling	120
		5.1.2 Nonuniform polyphase representation	121
	5.2	Shifting the frequency response	126
6	**Analysis of multiphase switched-capacitor networks**		**129**
	6.1	Circuit analysis	129
	6.2	Signal-flow graph analysis	134
		6.2.1 Signal-flow graph concepts	135
		6.2.2 Topological formulae	139
		6.2.3 SSN-type SC networks	144
		6.2.4 SSN-network transformation	156
		6.2.5 Analysis of general SC networks	163
7	**FIR switched-capacitor filters**		**169**
	7.1	Building-blocks for FIR SC filters	170
		7.1.1 Memory elements	171
		7.1.2 Multiplier and summer circuits	182
	7.2	Morphological design of FIR SC filters	186
		7.2.1 Basic FIR SC filter structures	186

				Contents	ix

		7.2.2	Composite FIR SC filter structures	192
		7.2.3	Evaluation of FIR SC filter structures	196
	7.3	Multirate FIR SC filters		199
		7.3.1	FIR SC decimators	199
		7.3.2	FIR SC interpolators	200

8	**IIR switched-capacitor filters**			**203**
	8.1	Double-frequency transformation		205
	8.2	VIS-SC circuits		211
		8.2.1	VIS-SC circuit concept	212
		8.2.2	Realization of VIS elements	214
		8.2.3	VIS-SC filter structures	217
	8.3	Transmission of effective pseudoenergy		221
		8.3.1	Losses of effective pseudoenergy	221
		8.3.2	VIS-SC circuits with recovery of effective pseudoenergy	224

9	**Applications of multirate and multiphase SC circuits**			**241**
	9.1	Transmultiplexers and highly selective filters		241
		9.1.1	Single-way transmultiplexers	241
		9.1.2	N-path and pseudo-N-path filters	242
		9.1.3	Frequency translated SC filters	243
	9.2	Data converters		244
		9.2.1	Oversampling converters	245
		9.2.2	Polyphase converters	246
	9.3	Adaptive filters		247
		9.3.1	Adaptive FIR filters	248
		9.3.2	Adaptive IIR filters	248

References	251
Index	271

Preface

Switched-capacitor circuits, or SC circuits for short, contain only switches, capacitors and active elements (e.g. operational amplifiers or unity-gain buffers). Such circuits are suitable for integration using MOS (metal-oxide-semiconductor) or GaAs (gallium arsenide) technology, because their functional features depend not on absolute capacitance values but on capacitance ratios, which can be precisely controlled in a technological process.

SC circuits are used in communications, measurement, control and related areas of technology for various signal processing applications, notably the realization of filters and filter banks but also for analog-to-digital and digital-to-analog data converters, oscillators, phase-locked loops, modulators, comparators, codecs, correctors, adaptive systems, and so on.

They belong to the class of analog sampled-data systems. Signals in SC circuits are usually sampled-and-held, i.e. they have a stepwise waveform. The stepwidth is called the sampling period and its inverse is referred to as the sampling rate.

Circuits discussed in this book are multirate in the sense that they process signals with various sampling rates or their signals are composed of sequences of various sampling periods. In the first case we deal with sampling rate alteration, i.e. increase (interpolation) or decrease (decimation). The second case corresponds in turn to nonuniform periodic sampling, referred to as heteromerous sampling. Both processes are actually particular cases of a general product modulation. Multirate signal processing is discussed in this generalized sense.

Multirate techniques were developed for more than 15 years ago, but recently they have found many applications in digital audio, speech and image data compression, data conversion, adaptive signal processing, highly selective filtering and many other fields.

Most SC circuits described in the literature are the so-called stray insensitive circuits, i.e. circuits insensitive to parasitic capacitances which are unavoidable in contemporary technology. It is also usually assumed that SC circuits are driven by a symmetrical two-phase clock, i.e. a clock corresponding to two switching phases: the even phase and the odd phase. Since new technologies, such as GaAs, relax the restriction of strict parasitic insensitivity (while admittedly imposing others) and multirate SC circuits are inherently multiphase, i.e. driven by a multiphase clock, general SC circuits are considered in this book.

The book is intended for technically sophisticated readers, i.e. advanced electrical engineering students and industry engineers working in the area of signal processing and circuit design. Thus, some background knowledge of university mathematics, circuit theory and electronics is desirable if not necessary for studying this book. For such readers the text should be self-contained.

An attempt has been made to distinguish fundamental and permanent concepts in the analysis and design of SC circuits from those which have appeared ephemeral or less important. Tending to this end, the text focuses on various FIR (finite impulse response) filter structures and on IIR (infinite impulse response) SC filters which simulate classical lossless circuits (e.g. doubly resistively terminated LC ladder filters) including the so-called pseudolossless (or more general pseudopassive) SC circuits and especially multirate SC circuits with recovery of the effective pseudoenergy. The book discusses also other promising approaches to the design of SC circuits. Special attention has been paid to the analysis of multirate and multiphase SC circuits using signal-flow graphs (SFGs).

Most of the material contained in this book is spread around in many papers, conference proceedings, doctorial dissertations, internal reports, etc., but some substantial information is new and has not been published before.

The book consists of nine chapters.

Chapter 1 provides an introduction and overview of the basic properties and applications of modern electronic filters, switched-capacitor circuits and multirate systems.

Chapter 2 gives a short introduction to the technological aspects of the design of SC circuits and describes (without detail) implementation of their basic elements in CMOS VLSI technology.

Chapter 3 is devoted to the description and analysis of sampled-data signals and systems. It presents also more advanced topics necessary to understand further parts of this book.

In Chapter 4, attention is focused on concepts of multirate signal processing and on the theory and realization of multirate systems. Various useful structures including polyphase decomposition, subband coding schemes, quadrature-mirror filter (QMF) banks and multirate wave arrangements are discussed.

Chapter 5 describes systems suitable for the realization of SC circuits with nonuniformly sampled signals. Their application for the precise shifting of the filter frequency response is presented.

Analysis of multirate and multiphase SC circuits, with special emphasis on the signal-flow graph approach, is the subject of Chapter 6. Methods for the derivation of network transfer functions in a closed (symbolic) form are provided.

Chapter 7 is fully devoted to the design of FIR SC filters. Basic building-blocks for these filters are described and the morphological approach to the derivation of various filter structures is discussed. Multirate FIR SC filter structures are also considered.

Chapter 8 covers the description of chosen IIR SC filters. First, a quite general method is presented for the design of SC circuits which simulate, with bilinear precision, voltages and charges in their continuous-time counterparts (called the reference circuits). This method is based on the double-frequency transformation. Then voltage inverter switch SC (VIS-SC) circuits which, in turn, simulate voltage waves in their reference circuits, are described. Multirate VIS-SC filter structures with recovery of the effective pseudo-energy are derived. They are characterized by substantial advantages compared to classical circuits.

Finally, chosen applications of multirate and multiphase SC circuits are discussed in Chapter 9, including transmultiplexers, highly selective systems, N-path and pseudo-N-path filters, data converters, sigma-delta modulators and adaptive filters.

The author wishes to acknowledge a number of people whose results appear in the text. Special thanks are addressed to Professors Alfred Fettweis, George Moschytz, Peter Noll, Joos Vandewalle, Zdzisław Korzec and Andrzej Napieralski, who have contributed in a variety of ways to the results presented in this book. The author is also very grateful to his colleagues Ulrich Menzi, Paul Zbinden, August Kaelin, Dieter Brückmann and Zygmunt Ciota for their help with preparing experiments, computer simulations, collecting illustrative materials and for many stimulating discussions. Many thanks should also go to my assistants, who were the first readers of this book.

Most of the presented results of the computer simulations and the experimental investigations were carried out by the author at the Lehrstuhl für Nachrichtentechnik, Ruhr-Universität Bochum, Germany, under a grant from the Alexander von Humboldt Foundation, and in the Institut für Signal- und Informationsverarbeitung, Eidgenössische Technische Hochschule Zürich, Switzerland. SCADAP chip, described in Chapter 9, was fabricated by EM-Microelectronic Marin, Switzerland.

My sincere gratitude goes to my parents: to my mother, whose good advice has helped me not only throughout my whole life but also in this project, and to my father, who was my first teacher in electrical engineering.

I would also like to express my deepest gratitude to my wife Agata, who showed love, patience and understanding during this long work and to my 13-year-old son Marcin, who prepared, with a computer, many figures in this book . Finally, I would like to thank my dog Intel, who helped me, as a result of the regular walks, to distinguish between day and night.

One day, during my deep absorption in writing this book, I was suddenly asked by my wife, who tried but could not keep its title in mind, what actually are multirate and multiphase switched-capacitor circuits and what are their applications. I noticed, to my great surprise, that is not easy to explain in few words what this book is about. I have tried to do that in this preface, but perhaps the best way is to read the book.

Adam Dąbrowski

1
Introduction

1.1 MODERN VERSUS CLASSICAL FILTERS

Electronic circuits for signal processing, and particularly those for filtering of signals, are subject to severe requirements for the precision of their elements, and that in contradistinction to typical electronic circuits for many other applications. Very precise elements (some of them possibly tunable) are usually required even for circuits with attainable small sensitivities to changes of parameters of their elements. That is why, in spite of the tremendous development of electronic technologies, which has enabled a continuous modernization and miniaturization of most electronic circuits, realizations of electric filters as passive LC ladders (but lossless at least under ideal conditions) connected between resistive terminations have remained practically constant for astonishingly many years – over some decades from the first realizations at the beginning of our century until the sixties. This is due to the excellent properties of such structures. They are always (structurally) stable and possess very low sensitivity to element parameter variations, if properly designed. The stability is guaranteed by the passivity because the total energy stored in a passive system plays the role of the Liapunov function (Balabanian and Bickart, 1983).

Stopband low sensitivity follows from the ladder structure. In this case, positions of the transmission zeros (frequencies of infinitely large attenuation) are determined separately by particular resonant circuits which can be individually tuned.

Finally, the low passband sensitivity is caused by the filter losslessness. This fundamental property, discovered independently by Fettweis (1960) and Orchard (1966), is illustrated in Fig. 1.1. If

2 Introduction

a lossless filter is properly designed, there exist passband frequencies (e.g. the frequency f_o) at which the attenuation vanishes to zero (Fig. 1.1b) and then the reflected power is also equal to zero (Fig. 1.1c). For a lossless system, the transmitted power cannot be greater than the incident power. Therefore, if any filter parameter x changes in value from the optimal value, say x_o, then the reflected power can only be positive and thus the filter attenuation can only rise (cf. the right-hand side of Fig. 1.1b). Hence, for the optimum parameter value x_o, the attenuation reaches a minimum and its sensitivity to changes of x for $x = x_o$ is zero. Since filter properties change smoothly, we conclude that the attenuation sensitivity to changes of parameter x stays low in the whole passband. Furthermore, since x is the value of an arbitrary filter element (except for terminating resistors), the filter overall passband sensitivity to element variations is low.

Although LC doubly resistively terminated ladder filters worked perfectly, they were large, heavy and expensive. Direct miniaturization of them was usually not possible because they had to carry – as passive networks – the whole effective signal energy to be transmitted. Inductances and capacitances of filter components follow directly from requirements for operational frequency range and for loading resistances. If we changed the loading resistances (using, for example, transformers or reducing the transmitted energy), then, for a particular frequency band, minimization of capacitances would lead to maximization of inductances, and vice versa. Thus, the size of a passive filter cannot be easily reduced.

For more than 30 years, big efforts have been made to integrate filters or at least to diminish their size, weight, power consumption and cost. New phenomena such as various semiconductor effects, the piezoelectric effect, the surface acoustic wave effect, etc., have been exploited. New components, e.g. controlled sources, operational amplifiers, buffers, gyrators, integrators, etc., have been used. Entirely new approaches such as analog or digital sampled-data filters have been proposed. Many of these new ideas have, however, been based on the concept of possibly exact simulation of classical LC filters. This is because of the excellent properties of such filters,

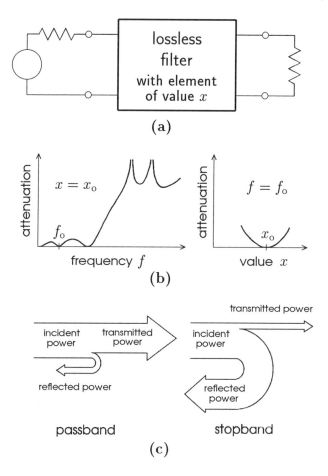

Fig. 1.1 Illustration of low passband sensitivity of classical lossless filters: (a) lossless filter with a chosen element of value x, (b) filter attenuation characteristics, (c) power transmission.

as discussed above.

The most cumbersome elements in LC filters are inductors. They are bulky, heavy, expensive and lossy. Elimination of inductors led to continuous-time active-RC filters which for small size were realized as hybrid integrated circuits. Such circuits contained monolithic integrated operational amplifiers (op amps) and miniature capacitors which were soldered to resistors printed on a plate. The printed resistors could even be tuned by a laser.

4 *Introduction*

Further reduction of the size of active-RC filters, e.g. their integration in a form of a single monolithic circuit, is, however, hardly possible. Since filter time constants have the dimension of an RC product, minimization of capacitances would lead to the maximization of resistances, and vice versa. Moreover, monolithic resistors are very imprecise, nonlinear, temperature dependent, restricted to a narrow range of values, and they cannot be tuned. All these drawbacks practically preclude integration of active-RC filters in a monolithic form.

The way for miniaturization and integration of electronic filters is in the application of low-power and low-voltage active circuits which do not need to carry the whole effective energy at the system output. This can be achieved either by means of precise analog active continuous-time or sampled-data circuits (in which a major part of the effective signal energy is drawn from the power supplies) or by means of digital circuits which process signals coded in a form of sequences of numbers, so their 'physical energy' plays a secondary role.

In analog or digital sampled-data systems, the physical energy of signals, or, in other words, the energy of physical representations of sequences of signal samples, may be related to the effective 'symbolic' signal energy in a quite complicated manner. Nevertheless, a notion of some energy-type function of the discrete time, directly related to the effective signal energy, can usually be introduced. Such a function (proposed first in Fettweis, 1972) is called in this book the **pseudoenergy**. In some so-called **pseudopassive** systems[*], this very useful function can play the role of the Liapunov function, and thus can guarantee system stability. If we simulate doubly resistively terminated LC filters element-by-element by pseudopassive or even pseudolossless sampled-data systems, then the marvellous properties of structural (robust) stability and low sensitivity of LC filters will be preserved also in these new designs. The most important systems in this class are **wave digital filters** (WDFs) and **voltage inverter switch switched-capacitor circuits** (VIS-SC

---

[*]Systems passive (lossless) with respect to the pseudoenergy function are called pseudopassive (pseudolossless).

circuits for short) proposed by Fettweis (1971, 1979a, b). VIS-SC filters belong to a broader class of so-called **switched-capacitor circuits** (or SC circuits), which are the subject of this book.

1.2 SWITCHED-CAPACITOR FILTERS

The revolution in the integration of electronic filters began in the seventies with achievements in MOS VLSI (metal-oxide-semiconductor very large scale integration) technology.

On the one hand, the ease of integrating perfect switches, precise capacitors (or more rigorously – precise capacitance ratios), and good op amps led to an entirely new class of analog circuits, namely, to the **switched-capacitor circuits** (or SC circuits for short) containing switches, capacitors and active elements. Multirate and multiphase versions of SC circuits are discussed in this book. The possibility of replacing resistors by configurations of switches and capacitors made the use of SC circuits as active analog filters very attractive*. SC filters were, in fact, the first general-purpose electronic filters successfully realized in the form of a single monolithic integrated circuit (Caves et al., 1977; Hostička, Brodersen and Gray, 1977).

On the other hand, digital single-chip integrated filters were developed concurrently with SC filters and the milestones on this way are general-purpose digital signal processors (DSPs) and application-specific integrated circuits (ASICs). About the seventies, digital systems began to replace analog solutions in almost all signal processing applications. One might have even thought that the era of classical frequency-domain signal processing, i.e. of high-precision analog filters, would have finished. This is, however, not true. In fact, digital systems generated entirely new and very important filtering tasks. Anti-aliasing and anti-imaging filters, as well as filters for multirate systems and filter banks, can serve as good examples. All these new filter classes can efficiently be realized with SC techniques and/or with a so-called mixed-mode approach, i.e. with

---

*This is, as we will see later on, due to the dependence of SC transfer functions on capacitance ratios which can be controlled very precisely in a technological process.

6   *Introduction*

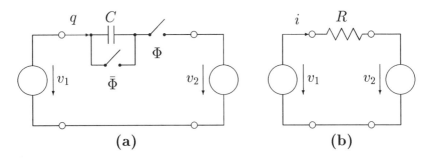

Fig. 1.2 Series switched-capacitor realization of a resistor: (a) SC circuit, (b) equivalent resistor.

partly analog (SC) and partly digital circuitry. The latter realizations were possible as a result of recent improvements in CMOS (complementary metal-oxide-semiconductor) and BiCMOS (bipolar/CMOS) technology. Especially attractive is the realization of an analog signal-processing unit and an analog interface on the same chip along with digital circuitry (Soin, Morris and Maloberti, 1991). This makes it possible to implement entire systems on single chips and provides the designer with great flexibility for partitioning the signal-processing tasks between analog and digital realizations, resulting in superior technical solutions. The single-chip system can possess a smaller number of components, simpler structure, and in consequence is characterized by smaller power consumption, smaller size, smaller weight, etc. This leads to reduced invention, design and production costs. Such systems are therefore particularly suitable for mass-production of portable and mobile equipment.

The digital part can also help to improve precision in the performance of analog circuitry by adding the possibility for complex control and compensation mechanisms. By this means we can also realize more advanced, e.g. analog, adaptive systems (Dąbrowski, 1993).

We shall now study the fundamental properties of SC circuits. As already mentioned above, configurations of switches and capacitors can simulate resistors. A simple example of an SC circuit which is equivalent to a resistor is shown in Fig. 1.2. Assume that the two-phase clock Φ, $\bar{\Phi}$ is applied with pulses occuring every T seconds.

Table 1.1 Typical data of MOS technology for SC circuits

Element	Area ($10^3 \mu m^2$)	Absolute tolerance (%)	Relative tolerance (%)	Temperature coefficient (ppm/K)	Voltage coefficient (ppm/V)
Op amp	20–100	—	—	—	—
Switch	1–10	—	—	—	—
Capacitor	2/(1pF)	10	0.05	10–50	20–200

We observe voltages v_1 and v_2 and the total charge q that has flowed through the circuit until instant $t_n = nT$. The charge increment $\triangle q(t_n)$ can then be expressed as

$$\triangle q(t_n) = q(t_n) - q(t_n - T) = C[v_1(t_n) - v_2(t_n)] \qquad (1.1)$$

and thus, the effective current as

$$i(t_n) = \frac{\triangle q(t_n)}{T} = \frac{C}{T}[v_1(t_n) - v_2(t_n)] = R^{-1}[v_1(t_n) - v_2(t_n)] \ . \qquad (1.2)$$

Hence, for the equivalent resistance R we obtain

$$R = \frac{T}{C} \ . \qquad (1.3)$$

From equation (1.3) we immediately conclude that time constants of SC circuits have the dimension of ratio (not product like RC type products in active-RC filters!) of capacitances times the operation period. This is a fundamental property of SC circuits which made their implementation in MOS technology, in opposition to active-RC filters, so successful. Indeed, the circuit capacitances may be freely scaled (without any influence on the time constants), and thus they may be straightforwardly reduced to the range of some tens of picofarads. Moreover, capacitor ratios may be controlled very exactly, i.e. up to ca. 0.05% , although the absolute capacitance values will be realized with typical tolerances of ca. 10% (cf. Table 1.1).

SC circuits belong to a general class of analog sampled-data systems. Other types of MOS integrated circuits in this class are, for example, charge-coupled devices (CCDs), charge-transfer devices

8 *Introduction*

(CTDs) and switched-current (SI) circuits. All of them are, however, restricted to particular filter structures and/or to particular applications. Thus, SC filters are usually more flexible and accurate than other analog sampled-data realizations, making them very attractive for integration as ASICs.

The need for constant miniaturization of electronic filters and for long battery life (e.g. in portable units such as mobile phones) makes low-power and low-voltage design criteria more and more important for both analog and digital integrated circuits. Smaller dimensions and higher densities reduce the isolation barriers to a few volts and the power preventing the chip from overheating to some milliwatts. For certain (in particular low-power and low-voltage) applications, switched-capacitor circuits have distinct advantages over their all-digital counterparts. The most important among them are:

- lower power consumption by about two orders of magnitude for applications in which a low dynamic range can be tolerated;

- simpler structure, resulting in lower inventory, design and production costs;

- larger degree of parallelism in the realization of certain signal-processing algorithms.

It follows from the above properties that SC circuits, compared with digital realizations based on currently available DSPs, can be preferable for processing signals at relatively high frequencies (i.e. in the range of several hundred kilohertz), thus also for multirate systems and filter banks, i.e. for applications which are discussed in this book. As an illustrative example, the performance limits for an FIR (finite impulse response) filter realized using SC and DSP techniques are qualitatively compared in Fig. 1.3. For the DSP realization we assume a single general-purpose DSP, e.g. a TMS320C5x processor. For a switched-capacitor realization, the filter length is limited by the maximum chip area and by the minimum signal-to-noise ratio (SNR) requirements. The upper limit of the sampling frequency, i.e. the frequency range of SC filters, is determined by the bandwidth of the op amps and by the maximum

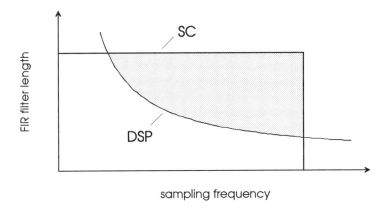

Fig. 1.3 Performance limits for SC and DSP techniques.

$\tau \approx RC$, where R is the on-resistance of the switches and C the maximum effective capacitance in the circuit which is in turn governed by the technology used. Using a CMOS process, frequencies of several hundred kilohertz can be achieved using clock rates of up to several megahertz. For higher frequencies a faster process should be used, e.g. gallium arsenide (GaAs) technology. For a DSP approach, the maximum signal frequency depends on the operation cycle duration. For FIR filters of very small length, this maximum frequency limit is, as a rule, higher than that for an SC realization. Unfortunately, the required filter lengths are usually too large to be able to utilize this advantage. Thus, for many applications requiring large filter lengths, the sampling rate of the DSP must be relatively low. Accordingly, for typical sampling rates below, say of 1 MHz, and filter lengths of the order of 50, the limits of realizability clearly favour the SC implementation, as confirmed in Fig. 1.3.

Another important criterion for the comparison of SC and digital VLSI filters is based on the interdependence between power consumption per filter pole and dynamic range (Vittoz, 1990). As shown in Fig. 1.4, SC circuits have a significant advantage over digital circuits in applications in which the dynamic range is low.

10   *Introduction*

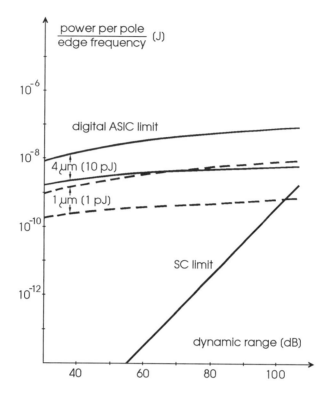

Fig. 1.4 Power per pole versus dynamic range for digital and SC VLSI filters.

An important filter performance criterion is the attainable transfer function Q-factor, which is measure of the filter selectivity, i.e. the reverse of the relative passband width. The approximate performance capabilities which can be achieved for the most important filter types are illustrated in Fig. 1.5. The vertical axis shows the attainable Q-factor and the horizontal axis the achievable operating frequency range. Typical SC filters are characterized by Q-factors limited to about 100 due to a limit on the maximum capacitance ratio (the maximum capacitance is limited by the maximum realizable chip and the minimum capacitance by the technological tolerance and the requirements for the minimum SNR). Attainable Q-factor can, however, be extended to some thousands using the multirate techniques described in this book. On the other hand, the frequency range limit of several hundreds of kilohertz can be extended to about

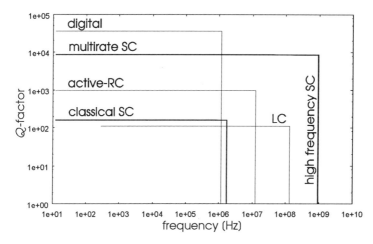

Fig. 1.5 Transfer function Q-factor and frequency range limits for SC, digital, active-RC and LC filters.

100 MHz by replacing CMOS technology by GaAs technology and also by the use of fast-settling op amps.

For comparison, classical LC filters can attain Q-factors of about 100 and a frequency range of the order of 100 MHz; but below say 1 kHz they become bulky and cumbersome. Active-RC filters can be designed for Q-factors up to about 1000 and for frequencies to around 10 MHz. Digital filters can achieve quite high Q-factors but they can operate up to only about 1 MHz.

As we can see, particularly for many low-power, low-voltage and/or low-dynamic range applications, as well as for high Q values and high-frequency applications including multirate systems and filter banks, SC realizations can be clearly preferable to other analog approaches and to all-digital approaches. Moreover, SC circuits will dominate in the implementation of the interface circuitry between digital systems and the external analog world.

Most of the SC circuits described in the literature are the so-called stray insensitive circuits (i.e. insensitive to parasitic capacitance which is unavoidable in current MOS technology). It is also usually assumed that SC circuits are driven by a symmetrical two-phase clock. Since new technologies, such as GaAs, relax the restriction of parasitic insensitivity and moderate the number of clock phases (while admittedly imposing others), not only multiphase but also general multirate SC circuits with, say, unity-gain buffer amplifiers instead of conventional op amps, may be expected in the near future (Haigh et al., 1991). This was the motivation for extending the analysis and design methods in this book beyond the conventional two-phase SC networks so as to cover the general multiphase SC networks, and thus, multirate systems including also filter banks.

1.3 MULTIRATE SYSTEMS

The notion of **multirate systems** is usually restricted to systems for sampling rate alteration (increase or decrease) (Crochiere and Rabiner, 1983; Vetterli and Herley, 1992; Fliege, 1993; Vaidyanathan, 1993). The corresponding area of science and technology is called **multirate signal processing**.

The usefulness of sampling rate alteration follows directly from the classical sampling theorem (Whittaker, 1915; Nyquist, 1928; Whittaker, 1929; Kotelnikov, 1933; Shannon, 1949) which connects the minimum required sampling rate with the signal spectrum bandwidth.* For simplicity of signal processing, minimization of sampling rate is reasonable in each signal processing stage, realized by fitting the sampling rate to the actual signal spectrum bandwidth. Indeed, if we modify the signal bandwidth by filtering, then, for simplicity of further signal processing, we have to reduce the sampling rate. Such a process is referred to as the **decimation**. The opposite process is called **interpolation**.

Limitation of multirate signal processing to sampling rate alteration only (i.e. decimation and interpolation) is, however, the traditional and, in the author's opinion, narrow sense of a much broader discipline. In this book, we use the notion of multirate systems in a more general sense. It is proposed to interpret multirate signal processing as a process of cascading product modulation

---

*This famous theorem states, in its simplest but most popular form, that the sampling rate must be at least twice as high as the maximum signal spectrum frequency.

with filtering.

M-fold sampling rate compression (or expansion), M being an arbitrary integer called in this book the **multirate period**, can be realized by means of a product modulation with the carrier signal consisting of consecutive M-element sequences of the form

$$1, \underbrace{0, 0, \ldots, 0}_{M-1 \text{ zeros}}$$

Generally, the most interesting, especially for SC circuits, are carrier sequences containing only values 0 and ±1. In such cases no multiplications are required. For sequences containing values 0 and 1, the product modulation is equivalent to the process of **heteromerous sampling***.

Thus, sampling rate alteration, nonuniform sampling and specific modulations can be considered as particular cases of this general process. Theory, analysis and design of switched-capacitor realizations of multirate and multiphase circuits considered in this book constitute an important part of this discipline.

Multirate systems find applications in many fields. In communications they are, for example, used for data format conversion (transmultiplexing), data transmission and compression, and adaptive and statistical signal processing. They also have applications in medicine, manufacturing, robotics, control, geophysics, underwater acoustics and military areas such as object/target recognition and identification and image processing. In applied mathematics they are used for numerical solution of differential equations. In signal theory they find applications in time-frequency and time-scale signal representation such as the short-time Fourier transform and the wavelet (multiresolution) transform (Chui, 1992; Cohen, 1995). Both transform types are useful for the analysis of nonstationary signals.

Nonuniform (e.g. **heteromerous**) sampling is typically used in radar and sonar signal processing. Recently it has been shown that by this means we can avoid problems with numerical instability during computations of multiresolution coefficients in wavelet analysis

---

*The term 'heteromerous sampling' is introduced by the author to describe the sampling process with nonuniformly but periodically spaced samples (Dąbrowski, 1981).

(Vaidyanathan, 1995). Heteromerous sampling can also help to stabilize numerical solutions of stiff and oscillatory differential equations (Dąbrowski, 1981).

The most important multirate systems are the so-called **filter banks**, i.e. sets of many filters, usually with characteristics translated in such a way that their passbands cover the whole frequency axis. The analysis filter bank decomposes a signal spectrum in a number of directly adjacent frequency bands. Then the synthesis filter bank recombines these spectra and reconstructs the original signal.

Realization of multirate systems using switched-capacitor techniques requires multiphase clocks. Such SC systems are called **multiphase switched-capacitor circuits**. The theory, analysis and design of multiphase switched-capacitor circuits, with emphasis towards multirate applications, are the subjects discussed in this book.

Switched-capacitor interpolators and decimators were introduced in the eighties (Gregorian and Nicholson, 1980; Ghaderi, Temes and Law, 1981; Grünigen, Brugger and Moschytz, 1981; Fettweis, 1982; Grünigen *et al.*, 1982; Franca, 1984). Since then SC multirate and multiphase systems have been systematically studied and a variety of new SC circuit concepts have been proposed (Dąbrowski, 1982a; Franca, 1985; Dąbrowski, 1988a; Franca and Santos, 1988; Dąbrowski, Menzi and Moschytz, 1989, 1992; Franca and Martins, 1990).

SC multirate systems successfully compete with all-digital realizations in many aspects of signal processing. There exist applications whose realization in VLSI circuits is possible only with an analog approach, e.g. using multirate SC techniques (Martins and Franca, 1989, 1991; Berg *et al.*, 1994; Uehara and Gray, 1994). There also exists a need to solve signal-processing problems using both analog and digital circuitry in a single mixed analog-digital integrated circuit realizing all signal-processing tasks in a single chip (Soin, Morris and Maloberti, 1991).

2
Basic elements of SC circuits

Switched-capacitor circuits contain switches (transmission gates), capacitors and op amps (or other active elements) which are referred to as their basic elements. Integrators, adders, inverters, delays, etc., are built with these basic elements. SC circuits may also contain additional building-blocks such as multipliers, sigma-delta modulators and even auxiliary digital circuits. Usually a timing circuit (a clock) is also a part of the SC circuit structure. The clock provides two or more nonoverlapping pulse sequences which control the switches.

The basic elements of SC circuits are shown in Fig. 2.1. When they are realized as components of an integrated circuit, parasitic capacitance is associated with them, as shown schematically on the right-hand side of Fig. 2.1. For an integrated capacitor, two parasitic capacitances exist: the bottom plate capacitance and the top plate capacitance, but the bottom plate capacitance is much greater and can achieve values of up to 10% of the capacitor value, depending on the technological process. The switch has two grounded parasitic capacitances between its main (switching) terminals and ground but also four non-grounded capacitances associated with its control terminals. The op amp has two parasitic capacitances from each input terminal to ground and a capacitance between both input terminals.

Parasitic capacitances are nonlinear, temperature dependent and they cannot be exactly controlled. Therefore, in some SC circuits referred to as **parasitic sensitive**, they are harmful and can seriously affect the switched-capacitor filter response. Fortunately, in almost all practical applications **parasitic compensated** (Fleischer, Ganesan and Laker, 1981) and even **parasitic insensitive**

16 Basic elements of SC circuits

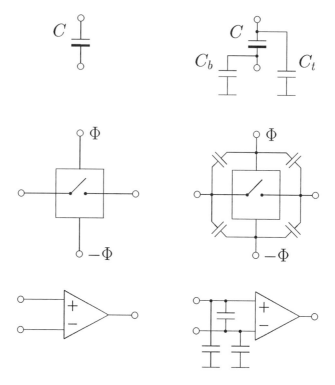

Fig. 2.1 Basic elements of SC circuits.

(Martin, 1980) SC circuits can be designed.

2.1 SWITCHES AND CLOCK

Switches transfer the charge within a switched-capacitor circuit, i.e. they charge and discharge capacitors. They are usually realized as transmission gates, i.e. as parallel connections of p-type and n-type transistors (Fig. 2.2). This complicates the switch structure, enlarges its area (one switch is realized with two transistors) and complicates its control (two complementary clock signals Φ and $-\Phi$ instead of one are necessary), but, on the other hand, it helps to remove the charge injection and also reduces the average value and variation of the switch on-resistance in the whole operating range of the drain-source voltage. Figure 2.3 shows that although the on-

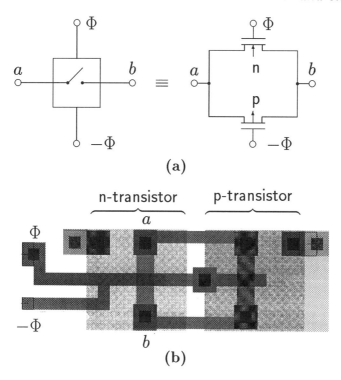

Fig. 2.2 Switch realized as a transmission gate: (a) circuit, (b) layout.

resistance of a single p-type and n-type transistor varies drastically as a function of the drain-source voltage, the transmission gate on-resistance varies quite moderately – between 10 kΩ and 25 kΩ only.

Integrated switches can be characterized by the following non-ideal effects:

- on-resistance R_{on}
- off-resistance R_{off}
- clock feedthrough
- noise.

Assume that a capacitor C is charged up through two switches (each of them associated with one of two capacitor plates), which

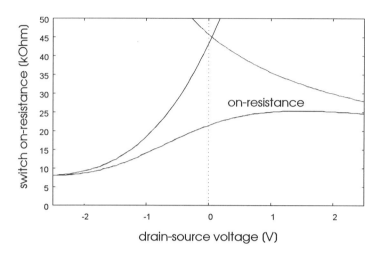

Fig. 2.3 The switch on-resistance as a function of drain-source voltage.

is a usual situation in SC circuits. Then, the charging process time constant is $2R_{on}C$. Taking into account that after approximately five time constants the capacitor C is practically fully charged, the charging condition can be formulated as

$$2R_{on}C \leq \frac{1}{5F}$$

where F is the sampling rate. Thus, for R_{on} we finally obtain

$$R_{on} \leq \frac{1}{10FC}. \quad (2.1)$$

For clock frequencies of say up to 1 MHz, typical capacitor values of the order of a few picofarads and $R_{on} \leq 30$ kΩ, requirement (2.1) is clearly fulfilled. Thus, the switch on-resistance becomes important only for high clock frequencies.

An additional effect caused by the non-zero switch R_{on} resistance is the noise generated by the switch. This effect is, however, usually so small that it can almost always be neglected.

A finite switch off-resistance R_{off} occurs due to the junction leakage effect. The reversely polarized p-n junction carries a small current which can erroneously discharge or charge particular capaci-

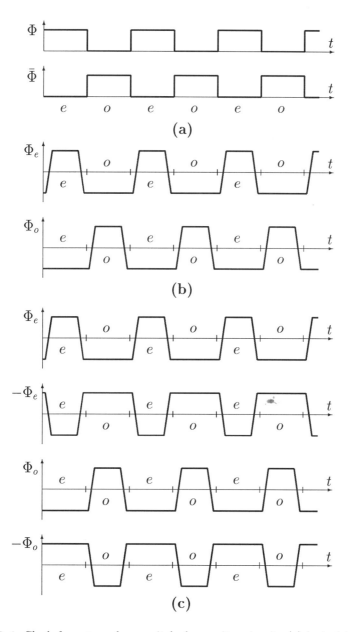

Fig. 2.4 Clock for a two-phase switched-capacitor circuit: (a) logical two-phase clock, (b) two-phase nonoverlapping clock, (c) practical four-phase clock corresponding to two-phase clocks in (a) and (b).

20 Basic elements of SC circuits

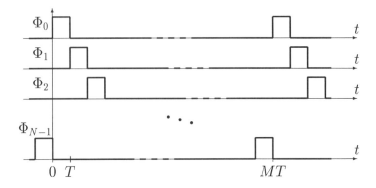

Fig. 2.5 A symmetrical (uniform) M-phase logical clock.

tors. This effect is also very small and for clock frequencies greater than several hundred hertz can be fully disregarded.

Clock feedthrough is a much more important effect. It consists in a cross-transmission of clock signal components into the signal path. This is due to the charge injection into the signal path as a transistor switches. In the result, spikes of the clock frequency occur and a strong DC offset can be present in the output signal. We can avoid these phenomena using a transmission gate with the same gate area for the p-type and n-type transistors. Then the clock feedthrough signals associated with both transistors will cancel each other. We can also reduce the gate area and enlarge the distance from the drain and the source to the gate, making the clock feedthrough effects smaller.

Switches are controlled by clock signals. In the simplest situation of the so-called two-phase SC circuit, two switch groups exist: even-phase switches (closed in even-phase instants) and odd-phase switches (closed in odd-phase instants). Theoretically, this situation can be described by one clock signal Φ (Fig. 2.4a) equal to logical 1 when even-phase switches are closed (odd-phase switches are open) and equal to logical 0 (the complementary signal $\bar{\Phi}$ equal to logical 1) in the opposite situation. In reality, however, a two-phase clock of two nonoverlapping signals Φ_e and Φ_o is necessary (Fig. 2.4b). For switches realized as transmission gates even a four-phase clock Φ_e, $-\Phi_e$, Φ_o and $-\Phi_o$ must be used (Fig. 2.4c). In most further considerations, we will use the simplified (logical) two-phase clock interpretation Φ, $\bar{\Phi}$ as illustrated in Fig. 2.4a.

Multirate and multiphase SC circuits require a multiphase clock. A symmetrical (uniform) M-phase logical clock for an M-phase SC circuit is shown in Fig. 2.5. In practice, we will of course need in this case a $2M$-phase real clock with pairwise opposite (complementary) phase signals, if switches are realized as transmission gates.

The pulse-width (or clock period) T of the clock in Fig. 2.5 corresponds to the network operation rate

$$F = \frac{1}{T}.$$

This can also serve as a sampling rate of signals at the circuit input and/or output. The input or output rate can, however, be smaller. It can even be M times less than F, i.e.

$$F_o = \frac{F}{M},$$

if the circuit input or output switches are controlled by a single-phase clock signal. Such a situation corresponds to an M-fold interpolator (cf. section 4.1.1) or to an M-fold decimator (cf. section 4.1.2), respectively. In both cases, the SC network operates with the higher sampling rate, i.e. with rate F. In many cases, however, we can decompose it into M so-called **polyphase components** (cf. section 4.4.4). Each of them operates with the lower sampling rate F_o and is driven by a respective single-phase component signal of the M-phase clock.

A multiphase clock in Fig. 2.5 is realized in practice on the basis of a high frequency system clock, say, with pulse-width (or period) T_c. The frequency

$$F_c = \frac{1}{T_c}$$

is then divided by some integer divider M_c in order to obtain the actual operation rate

$$F = \frac{F_c}{M_c}.$$

Changes of the value of this divider lead to nonuniform sampling. Such a technique is referred to in this book as **commensurate sampling**, because individual sampling periods are all multiples

22 *Basic elements of SC circuits*

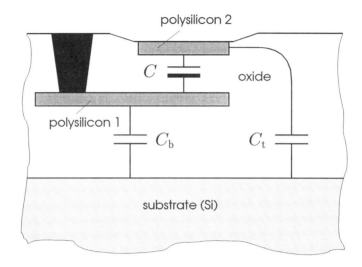

Fig. 2.6 An MOS capacitor.

of the system pulse-width T_c. If the stream of variable dividers is periodic, i.e. composed of the repetitive sequence of, say, M values $M_1, M_2, \ldots, M_M$, the resultant nonuniform periodic sampling process is called **heteromerous commensurate sampling** of order M. This technique is discussed in detail in Chapter 4.

2.2 CAPACITORS

An MOS capacitor is shown in Fig. 2.6. The top plate and the bottom plate are realized as polysilicon layers. They are separated by an insulating oxide layer. Between the bottom plate and the substrate (Si) a protective layer can additionally be made, in order to reduce the capacitive bottom plate coupling to the substrate. As already mentioned above, there are two parasitic capacitances C_t and C_b associated with the top plate and the bottom plate of the capacitor C, respectively. The bottom plate capacitance C_b is much higher than the top plate capacitance C_t and can even be as high as 10% of the desired capacitance C.

Since SC circuits cannot be tuned after fabrication, it is imperative that capacitor value ratios (which determine filtering characteristics) should be exact. For that reason, capacitors are realized as parallel connections of small unit capacitors and SC filter structures insensitive to parasitic capacitance are used. In order to obtain the exact capacitor value (e.g. between integer multiples of the unit capacitance), a single non-unit capacitor can additionally be connected in parallel to the configuration of unit capacitors.

The top plate shape of a unit capacitor should, on the one hand, be chosen to maximize the possible area to perimeter ratio, in order to reduce the fringe effects (effects associated with randomly nonsmooth top plate edges due to nonideal etching), and on the other hand, it should minimize the overall capacitor area. The optimum theoretical shape would be the regular hexagon but the suboptimum square shape can much more easily be realized in the integrated circuit and is used in practice (Gregorian, Martin and Temes, 1983).

A square unit capacitor can be characterized by the following parameters:

unit length	l	(μm)
unit area	$l \times l$	(μm^2)
main capacitance per unit area	$C_a/(l \times l)$	(pF/μm^2)
fringe capacitance per unit length	C_l/l	(pF/μm)

Therefore, the total unit capacitance C_u can be computed as

$$C_u = l^2 \times C_a/l^2 + 4l \times C_l/l = C_a + 4C_l . \tag{2.2}$$

The usefulness of the idea of unit capacitors can be shown by the following example. Consider the realization of two capacitors C_1 and C_2 that should be matched. Assume that

$$C_2 = \alpha C_1$$

but $C_1 = n^2 C_u$ and $C_2 = m^2 C_u$, i.e.

$$\alpha = (m/n)^2 .$$

Two possible layouts of capacitors C_1 and C_2, those without and with unit capacitors, are shown in Fig. 2.7. For the realization

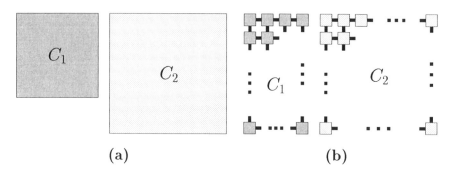

Fig. 2.7 Two possible layouts for matched capacitors C_1 and C_2: (a) layout without unit capacitors, (b) layout with unit capacitors.

without unit capacitors (Fig. 2.7a), we get

$$\alpha_1 = \frac{(ml)^2 \times C_a/l^2 + 4ml \times C_l/l}{(nl)^2 \times C_a/l^2 + 4nl \times C_l/l} = \frac{m^2 C_a + 4mC_l}{n^2 C_a + 4nC_l} \neq \alpha ,$$

i.e. the capacitors C_1 and C_2 are not exactly matched. Error in the ratio α_1 occurs due to the fringe capacitance, as the area to perimeter ratio has different values for capacitors C_1 and C_2. On the other hand, for the realization with unit capacitors (Fig. 2.7b), the area to perimeter ratio remains constant. Using (2.2) we compute

$$\alpha_2 = \frac{m^2(C_a + 4C_l)}{n^2(C_a + 4C_l)} = \left(\frac{m}{n}\right)^2 = \alpha$$

i.e. in this case the capacitors C_1 and C_2 are exactly matched.

As already mentioned above, for the realization of an arbitrary capacitor ratio, a single non-unit capacitor may be necessary. In order to keep the area to perimeter ratio constant, its value must be larger than one unit capacitance (but usually smaller than two). This can be achieved, for example, by using a rectangular but sufficiently elongated non-unit capacitor. By this means, however, very long capacitors can result, which are thus inefficient in area. To avoid long non-unit capacitors, notches or even holes can be made in the top plates to sufficiently increase the perimeter. Another technique consists in making stubs connected to unit capacitors (Fig. 2.8). A stub is half the width of one side of a unit capacitor. Its length determines the fraction of the unit capacitance to

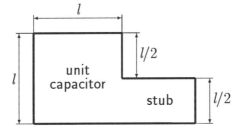

Fig. 2.8 Unit capacitor with a stub.

be added. Assume that the stub should realize the capacitance αC_u, $0 < \alpha < 1$. Thus the stub's length is

$$\frac{\alpha l^2 C_a}{(l/2)C_a} = 2\alpha l$$

and the total capacitance C_{us} of a unit capacitor connected with this stub is

$$C_{us} = \left[l^2 + (l/2)2\alpha l\right] C_a/l^2 + (4l + 2 \times 2\alpha l)C_l/l = (1+\alpha)(C_a + 4C_l) \ .$$

After substitution of (2.2), we finally get

$$C_{us} = (1+\alpha)C_u \ .$$

Hence, the effect of fringe capacitance has indeed been cancelled.

To improve matching of capacitors, they should be laid out close together, forming a so-called capacitor bank. Around the area of the complete capacitor bank, dummy capacitors can be placed. This improves the cancellation of the influence of fringe capacitance by keeping it constant, as the fringe capacitance is influenced by the adjacent capacitors. Therefore, by adding dummy capacitors, the fringe capacitance at the bank edge is the same as that within the bank. Dummy capacitors also help to improve the quality of etching of the top plates of unit capacitors. The etching of a one-unit capacitor top plate is affected by the adjacent top plate, and this can result in non-constant unit capacitor sizes, and thus may cause errors in the capacitor ratios. Therefore, the distance between the unit capacitors should be kept constant so that the etching effect is

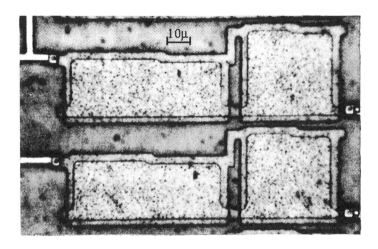

Fig. 2.9 A microphotograph of capacitors integrated in 2 μm CMOS technology.

constant. By placing dummy capacitors around the capacitor bank, the etching of all capacitors within the bank is the same.

Fringe capacitance near plate corners is different to that along plate edges. The error caused by this effect is, however, usually negligible.

The effects of fringe capacitance can be cancelled by using the above unit capacitor technique. Similarly, the influence of parasitic capacitance from the top and bottom plates (including metal and/or polysilicon interconnections) to the substrate (or ground) can be removed, as already mentioned, by the appropriate (parasitic-insensitive) structure. The only parasitic capacitance whose influence cannot be removed or compensated is the capacitance from the metal interconnections of the unit capacitor top plates to the capacitor bottom plate. This capacitance can, however, be minimized by appropriate layout (Taylor and Horvat, 1993). Its effect is, besides, quite small and becomes important only for high order and high Q (i.e. high selectivity) SC filters.

As an example, a microphotograph of a part of capacitors integrated in 2 μm CMOS technology is shown in Fig. 2.9 (Ciota, 1996). Two unit capacitors are visible on the right-hand side of Fig. 2.9 and two non-unit capacitors on the left-hand side.

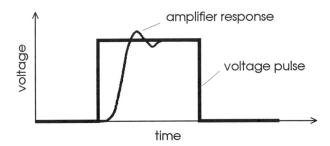

Fig. 2.10 Amplifier response in the time domain.

2.3 AMPLIFIERS

Operational amplifiers (op amps) are typical active elements in SC circuits but operational transconductance amplifiers (OTAs) are usually used instead of conventional (low output-impedance) op amps. Among important advantages of OTAs (compared to conventional op amps) are smaller layout area and lower power consumption.

Amplifiers limit the highest sampling rate and the filter (passband/stopband) edge frequencies (the amplifier bandwidth determines how high the clock frequency for the filter can be). The amplifier bandwidth determines also how large the capacitors can be. The speed of an amplifier (the bandwidth) is determined by the technological process. For example, gallium arsenide amplifiers are generally faster than their CMOS counterparts.

Therefore, two general requirements for amplifiers, to make them appropriate for SC circuits, can be formulated as follows:

- the DC gain must be greater than approximately 68 dB;

- the bandwidth must be at least four times larger than the highest (two-phase) clock frequency.

Finite amplifier gain results in the degradation of filter characteristics (lower Q factor, moved pole positions, higher passband ripple, broader transition bands).

28 *Basic elements of SC circuits*

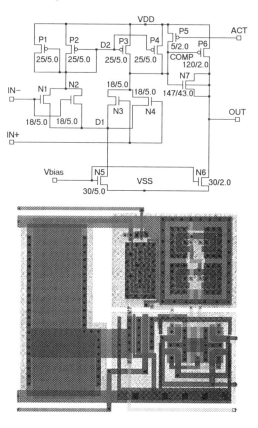

Fig. 2.11 Scheme and layout of the Miller OTA (op amp 1); size 185×170 μm.

Too narrow bandwidth can also degrade the filter characteristics because oscillatory transients may not have enough time to vanish. Therefore, it is essential to make sure that the amplifier can charge up its associated capacitors to a steady state with the prescribed precision (say 99%) within one clock pulse*. In practice, the amplifier should charge the associated capacitors in approximately half a clock pulse-width (Fig. 2.10) to guarantee the settling under worst

---

*Feedback capacitors (i.e. those connected from the amplifier output to input) tend to have great values. However, these are never completely discharged and the amplifier has only to fill up their charges in each phase. More important are those capacitors which are switched to ground (or to virtual ground) because they need to be completely charged and discharged in successive phases. Fortunately, they are usually small (a few picofarads).

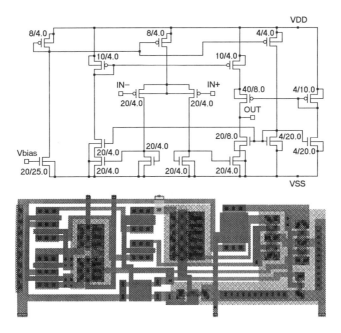

Fig. 2.12 Scheme and layout of the cascode OTA (op amp 2); size 270 × 90 μm.

case conditions. From this requirement we conclude that the amplifier bandwidth should be at least four times greater than the highest clock frequency (that corresponding to two clock pulses).

Another important drawback of integrated amplifiers is their input offset voltage. Offset is caused by the transistor mismatch and threshold. To reduce the offset, input transistors of the amplifier should be large (i.e. increased in width), as this improves their matching. By this means, they are affected by process gradients in approximately the same way. Offset effects can be usually compensated using special filter structures and appropriate switching.

Schemes and layouts of three typical OTAs used by the author in test circuits integrated in 2 μm CMOS technology by EM Microelectronic-Marin SA (Marin, Switzerland) are shown in Figs 2.11–2.13 (Zbinden and Dąbrowski, 1992; Dąbrowski and Zbinden, 1993; Zbinden, 1993). Their technical data are summarized in Table 2.1.

30 *Basic elements of SC circuits*

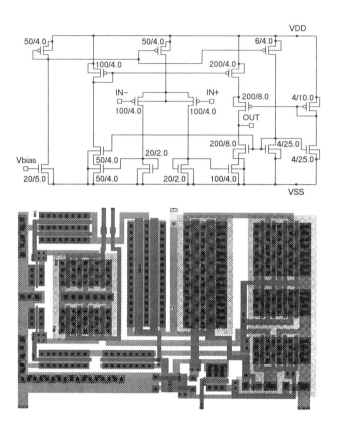

Fig. 2.13 Scheme and layout of the cascode OTA (op amp 3); size 290 × 180 μm.

Table 2.1 Technical data of illustrative OTAs for SC circuits

Amplifier	VDD	VSS	Bias current	Power consumption	DC gain	Gain-bandwidth product	Phase margin
	(V)		(μA)	(μW)	(dB)	(MHz)	(Deg)
op amp 1	+2.5	−2.5	5	170	69	0.92	55
op amp 2	+2.5	−2.5	4	100	> 85	3.7	55
op amp 3	+2.5	−2.5	20	450	> 85	5.5	60

3

Discrete-time systems

Switched-capacitor circuits operate as analog but discrete-time systems. In their signals, usually only step-wise changes in values occur (Fig. 3.1). A continuous-time step-wise signal can be fully described by a discrete-time signal, i.e. by a series of samples (Fig. 3.1). In this chapter we discuss chosen problems concerning discrete-time signals and systems in order to introduce the concepts and notions in this book, especially those which are used in a modified and/or generalized sense.

3.1 DISCRETE-TIME SIGNALS

3.1.1 Sampling process

Sampling of a continuous-time signal

$$x = x_c(t) , \quad -\infty < t < \infty \tag{3.1}$$

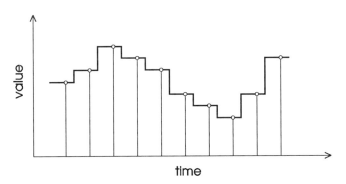

Fig. 3.1 Sampling of a step-wise signal.

is the process of time discretization. It consists in representing the signal $x_c(t)$ with a series of samples

$$x_n = x(t_n) = x_c(t_n) \; , \quad n = 0, \pm 1, \pm 2 \ldots \tag{3.2}$$

referred to as a **discrete-time signal**.

Uniform sampling with period $T > 0$ and rate

$$F = 1/T \tag{3.3}$$

is defined by

$$t_n = t_0 + nT \; , \quad n = 0, \pm 1, \pm 2 \ldots \tag{3.4}$$

where t_0 is an instant fulfilling the condition*

$$0 \leq t_0 < T \; . \tag{3.5}$$

We will also consider nonuniform periodic sampling consisting of repetitive sequences of different steplengths. Such a process is referred to in this book as **heteromerous sampling** (Dąbrowski, 1988b). This technique is widely used in the processing of radar and sonar signals (Wojtkiewicz, Tuszyński and Klimkiewicz, 1985). It has also been shown (Dąbrowski, 1981, 1982c, 1983) that heteromerous time discretization can help to stabilize computations for numerical simulation of dynamic systems. Recently, a similar effect has been shown for signal reconstruction after heteromerous sampling (Vaidyanathan, 1995; Vaidyanathan and Phoong, 1995).

The number of steps in a single sequence is called the order of the heteromerous sampling. The heteromerous sampling of Mth order can be interpreted as M interlaced uniform sampling processes, all with the same period equal to the sum of the sequence

---

*In most publications concerning discrete-time signal processing, it is assumed that $t_0 = 0$ and that the sampling period is normalized to $T = 1$. However, for SC circuits, especially for VIS-SC (voltage inverter switch SC) circuits and also for wave discrete-time systems, it is reasonable to assume 'physical' sampling instants, i.e. the general t_0 defined by equation (3.4) and the general T (Tsividis, 1983; Fettweis, 1984, 1986). A consequence of such an assumption is the appropriate modification of the Fourier discrete-time signal representation (section 3.1.3) and the $\mathcal{Z}$ transformation (section 3.1.5). Physical sampling instants should also be taken into account for the description of multirate systems and systems with nonuniformly sampled data (Chapters 4 and 5).

steplengths. This interpretation leads to the so-called **polyphase representation** of discrete-time signals, discussed in section 3.1.6. For example, for a heteromerous sampling of second order we have

$$T_1, T_2 > 0 \text{ and } T_1 + T_2 = 2T$$

which can be interpreted as two interlaced processes

$$x_{2n} = x(t_{2n}) = x(t_0 + 2nT) \tag{3.6}$$

$$x_{2n+1} = x(t_{2n+1}) = x(t_1 + 2nT) \tag{3.7}$$

with

$$t_1 = t_0 + T_1$$

and generally

$$t_{2n+1} = t_{2n} + T_1 \; , \quad n = 0, \pm 1, \pm 2, \ldots \; . \tag{3.8}$$

If, for example, $T_1 = T_2 = T$, then the second-order heteromerous sampling will be reduced to uniform sampling with period T.

3.1.2 Quantization

Quantization is the process of amplitude discretization, i.e. of converting a signal with continuous amplitude to a signal with discrete amplitude levels (Gersho, 1978). Sampling and quantization are necessary together for conversion of an analog signal into a digital form, the process usually realized with multirate switched-capacitor circuits (Maloberti, 1991).

The error due to the quantization process is referred to as quantization noise. This noise is usually analysed under the following assumptions:

- the quantization steps are uniform
- all quantization levels occur with equal probability
- the number of quantization levels is high
- the quantization noise is not correlated with the signal to be quantized.

34 Discrete-time systems

The first assumption is usually satisfactorily fulfilled if the resolution is not too high. If the steps are, however, not uniform, then the error will be a function of the input signal and thus it will not be an additive noise any more. The second assumption is almost never exactly fulfilled. Fortunately, if the number of quantization levels is large, which is almost always the case (the third assumption), then it will be satisfactorily fulfilled. The last assumption is perhaps the weakest, but any errors made due to it, e.g. in the calculation of the signal-to-noise ratio, can usually be neglected.

Denote by Q the quantization step and by $p(x)$ the probability distribution function of the quantization error. Then from the second assumption

$$\int_{-Q/2}^{Q/2} p(x)\,dx = 1 \tag{3.9}$$

where

$$p(x) = \begin{cases} \frac{1}{Q} & \text{for } x \in \left[-\frac{Q}{2}, \frac{Q}{2}\right] \\ 0 & \text{otherwise} \end{cases} \tag{3.10}$$

From equations (3.9) and (3.10), the average quantization error power N_Q can be calculated as

$$N_Q = \int_{-\infty}^{\infty} x^2 p(x)\,dx = \int_{-Q/2}^{Q/2} \frac{x^2}{Q}\,dx = \frac{Q^2}{12} \tag{3.11}$$

The signal-to-noise ratio in decibels is then

$$\text{SNR} = 10 \log_{10}\left(\frac{N_S}{N_Q}\right) \tag{3.12}$$

where N_S is the time-averaged signal power. Assume that the analog-to-digital converter (ADC) has a full scale of n bits. Then the maximum input signal amplitude is

$$A = (2^n - 1)Q$$

and thus

$$N_S \sim [(2^n - 1)Q]^2 \ .$$

From (3.12) it follows that

$$\text{SNR} \approx n 20 \log_{10} 2 + \text{const} \approx 6.02n + \text{const} \ . \tag{3.13}$$

3.1.3 Fourier representation of discrete-time signals

A continuous-time signal $x_c(t)$ has the following Fourier representation

$$X_c(j\omega) = \int_{-\infty}^{\infty} x_c(t) e^{-j\omega t} dt \; , \qquad (3.14a)$$

$$x_c(t) = \frac{1}{2\pi} \int_{-\infty}^{\infty} X_c(j\omega) e^{j\omega t} d\omega \; . \qquad (3.14b)$$

Expression (3.14a) is referred to as the Fourier transformation and expression (3.14b) as the inverse Fourier transformation.

The Fourier representation of a discrete-time signal $x(t_n)$ is defined by the discrete-time Fourier transformation

$$\begin{aligned} X\left(e^{j\omega T}\right) &= \sum_{n=-\infty}^{\infty} x(t_n) e^{-j\omega t_n} \\ &= \left(e^{j\omega T}\right)^{-t_0/T} \sum_{n=-\infty}^{\infty} x(t_n) \left(e^{j\omega T}\right)^{-n} \\ &= \left(e^{j\omega T}\right)^{-t_0/T} \mathcal{X}\left(e^{j\omega T}\right) \end{aligned} \qquad (3.15a)$$

where

$$\mathcal{X}\left(e^{j\omega T}\right) = \sum_{n=-\infty}^{\infty} x(t_n) \left(e^{j\omega T}\right)^{-n} \qquad (3.15b)$$

and by the inverse discrete-time Fourier transformation

$$x(t_n) = \frac{1}{\Omega} \int_{-\Omega/2}^{\Omega/2} X\left(e^{j\omega T}\right) e^{j\omega t_n} d\omega \qquad (3.15c)$$

where

$$\Omega = 2\pi/T = 2\pi F \; . \qquad (3.16)$$

Function $X(e^{j\omega T})$ is the modified Fourier transform* (spectrum) of the discrete-time signal $x(t_n)$ while function $\mathcal{X}(e^{j\omega T})$ is its conventional spectrum. Note that

$$\left|X\left(e^{j\omega T}\right)\right| = \left|\mathcal{X}\left(e^{j\omega T}\right)\right|$$

---

*Note that in expression (3.15a) we have assumed that $(e^{j\omega T})^{-t_0/T} = e^{-j\omega t_0}$, which means that the function $(z)^{-t_0/T}$ is considered in the Riemann surface of the function $\ln z$ (cf. section 3.1.5). This simply means that the argument of a complex value $(z)^{-t_0/T}$ is expressed in 'absolute' angular measure (i.e. using the whole real axis) instead of the normally used 'relative' measure restricted to an interval of length 2π, e.g. to the interval $(-\pi, \pi]$. In other words, the term $(e^{j\omega T})^{-t_0/T}$ should be considered as a function of $e^{j\omega t_0}$. Thus, on the Gaussian plane, the modified Fourier transform should be strictly denoted as $X(e^{j\omega T}, e^{j\omega t_0})$. The notation $X(e^{j\omega T})$ is, however, used for simplicity.

and for $t_0 = 0$ even

$$X\left(e^{j\omega T}\right) = \mathcal{X}\left(e^{j\omega T}\right) \ .$$

From equation (3.15b), we conclude that the conventional spectrum $\mathcal{X}(e^{j\omega T})$ is periodic on axis ω with period Ω. The modified spectrum $X(e^{j\omega T})$ is, in general, not periodic but the magnitude spectrum $|X(e^{j\omega T})|$ remains periodic.

Samples $x(t_n)$ can be computed not only from formula (3.15c) but also from expression (3.14b) simply by putting $t = t_n$. Comparing these results, we conclude that the continuous-time signal spectrum $X_c(j\omega)$ and the discrete-time signal spectrum $X(e^{j\omega T})$ are related as follows:

$$X\left(e^{j\omega T}\right) = \frac{1}{T} \sum_{k=-\infty}^{\infty} X_c(j\omega + jk\Omega)e^{jk\Omega t_0} \ . \tag{3.17a}$$

In the particular case for $t_0 = 0$, equation (3.17a) reduces to the famous form

$$X\left(e^{j\omega T}\right) = \mathcal{X}\left(e^{j\omega T}\right) = \frac{1}{T} \sum_{k=-\infty}^{\infty} X_c(j\omega + jk\Omega) \tag{3.17b}$$

which shows that the spectrum $\mathcal{X}(e^{j\omega T})$ is composed of replicas (in other words aliases or images) of the continuous-time signal spectrum $X_c(j\omega)$ moved along the ω axis by integer multiples of the sampling frequency Ω. This is illustrated in Fig. 3.2. If these replicas do overlap (Fig. 3.2b), the continuous-time signal $x_c(t)$ cannot be reconstructed from its samples $x(t_n)$ and we say that **nonrecoverable aliasing** occurs. To avoid it, the so-called **anti-aliasing** filter must precede the discrete-time signal processing system, in order to reduce the input signal spectrum. Then, this input signal can be reconstructed from its samples by means of the so-called **anti-imaging** filter, which suppresses all images (aliases) in spectrum $X(e^{j\omega T})$ and thus recovers the original spectrum $X_c(j\omega)$. This problem is discussed in detail in section 3.1.4.

Consider now a periodic signal $q(t_n)$ with period

$$T_o = MT \ , \tag{3.18}$$

Discrete-time signals

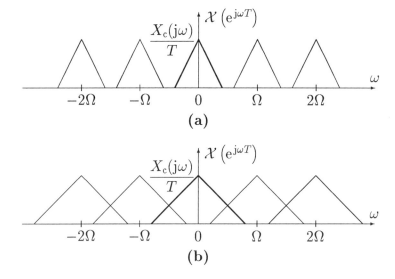

Fig. 3.2 Discrete-time signal spectrum: (a) nonrecoverable aliasing does not occur, (b) nonrecoverable aliasing occurs.

where $M \geq 1$ is an integer. The signal $q(t_n)$ has the following Fourier representation

$$Q(m\Omega_o) = \frac{e^{-jm\Omega_o t_0}}{M} \sum_{n=0}^{M-1} q(t_n) e^{-j2\pi mn/M}$$
$$= \frac{1}{M} \sum_{n=0}^{M-1} q(t_n) e^{-jm\Omega_o t_n} , \qquad (3.19a)$$
$$m = 0, 1, \ldots, M-1$$

and

$$q(t_n) = \sum_{m=0}^{M-1} Q(m\Omega_o) e^{jm\Omega_o t_0} e^{j2\pi mn/M}$$
$$= \sum_{m=0}^{M-1} Q(m\Omega_o) e^{jm\Omega_o t_n} , \qquad (3.19b)$$

where

$$\Omega_o = 2\pi/T_o = \Omega/M . \qquad (3.20)$$

Signal $q(t_n)$ can be explicitly described by an M-element sequence, e.g.

$$\tilde{q} = (q_0, q_1, \ldots, q_{M-1})$$

where
$$q_n = q(t_n) \ , \quad n = 0, 1, \ldots, M - 1 \ .$$

For a sequence $\tilde{q}$ the following Fourier representation exists

$$Q_m = \frac{1}{M} \sum_{n=0}^{M-1} q_n \mathrm{e}^{-\mathrm{j}2\pi mn/M} = \frac{1}{M} \sum_{n=0}^{M-1} q_n W_M^{mn} \ , \qquad (3.21\mathrm{a})$$
$$m = 0, 1, \ldots, M - 1$$

and

$$q_n = \sum_{m=0}^{M-1} Q_m \mathrm{e}^{\mathrm{j}2\pi mn/M} = \sum_{m=0}^{M-1} Q_m W_M^{-mn} \ , \qquad (3.21\mathrm{b})$$
$$n = 0, 1, \ldots, M - 1$$

where

$$W_M = \mathrm{e}^{-\mathrm{j}2\pi/M} \ . \qquad (3.22)$$

Expressions (3.21a) and (3.21b) are referred to as the DFT (discrete Fourier transformation) and IDFT (inverse DFT), respectively.

From expressions (3.19a) and (3.21a) we immediately obtain

$$Q(m\Omega_\mathrm{o}) = \mathrm{e}^{-\mathrm{j}m\Omega_\mathrm{o} t_\mathrm{o}} Q_m \ . \qquad (3.23)$$

In the literature, e.g. in (Oppenheim and Schafer, 1975) one can find a slightly different definition for DFT and IDFT: the factor $\frac{1}{M}$ occurs in IDFT instead of in DFT. In the author's opinion, however, the definition (3.21a) and (3.21b) is more convenient because Q_0 defined by expression (3.21a) for $m = 0$ is then directly the signal average (i.e. the DC component). This definition can also be found in the literature, e.g. in Azizi (1987). Another interesting approach would be a 'fair deal' of the coefficient among DFT and IDFT, i.e. the use of factor $\frac{1}{\sqrt{M}}$ in both transforms*.

3.1.4 Sampling theorems

Now we are interested in the reconstruction of the continuous-time signal $x_\mathrm{c}(t)$ on the basis of its samples, i.e. the discrete-time signal $x(t_n)$. Assume that the spectrum $X_\mathrm{c}(\mathrm{j}\omega)$ is restricted to the

---
*In such a case, the coefficient M disappears from the Parseval equation (3.64b), and thus both transforms are 'strictly' pseudolossless.

band $[-\Omega/2, \Omega/2]$ and does not contain Dirac pulses δ in points $\pm\Omega/2$. The corresponding signal $x_c(t)$ is referred in this book to as the **lowband signal** with respect to frequency Ω. In this case nonrecoverable aliasing does not occur and we can reconstruct the continuous-time signal spectrum $X_c(j\omega)$ using the anti-imaging filter with transfer function $T \operatorname{rect}(2\omega/\Omega)$. Indeed, from expression (3.17a) we obtain

$$X_c(j\omega) = T \operatorname{rect}\left(\frac{2\omega}{\Omega}\right) X\left(e^{j\omega T}\right), \qquad (3.24)$$

where

$$\operatorname{rect}(\omega) = \begin{cases} 1 & \text{for } |\omega| < 1 \\ 1/2 & \text{for } |\omega| = 1 \\ 0 & \text{for } |\omega| > 1 \end{cases} \qquad (3.25)$$

Hence, we can explicitly assign the continuous-time signal $x_c(t)$ to the given discrete-time signal $x(t_n)$ by means of the following reconstruction formula

$$x_c(t) = \sum_{n=-\infty}^{\infty} x(t_n) \operatorname{sinc}\left(\frac{\pi}{T}(t - t_n)\right) \qquad (3.26)$$

in which

$$\operatorname{sinc} t = \sin t / t . \qquad (3.27)$$

Reconstruction formula (3.26) is the basis for one of the most important theorems in the theory of discrete-time signal processing – the famous sampling theorem, which in its most elementary form can be formulated as follows:

> A continuous-time signal $x_c(t)$ can be reconstructed on the basis of the discrete-time signal $x(t_n)$ if the sampling rate F is at least twice as high as the greatest frequency in the continuous-time signal spectrum $X_c(j\omega)$ or, in other words, if the whole signal spectrum lies below $F/2$ called the **Nyquist frequency**.

This theorem was discovered by Whittakers (1915, 1929) but was brought into engineering by Nyquist (1928), Kotelnikov (1933) and Shannon (1949). There exist many extensions and generalizations of this theorem (Jerri, 1977; Crochiere and Rabiner, 1983; Dąbrowski, 1988b; Zayed, 1993). One of the most important is the following theorem formulated by Kohlenberg (1953):

Signal $x_c(t)$ of real variable t, with spectrum reduced on axis $\omega > 0$ to the band

$$\omega \in [\omega_o, \omega_o + \Omega/2] \ , \ \omega_o, \Omega \geq 0$$

which in points ω_o, $\omega_o + \Omega/2$ does not contain Dirac pulses δ, can be uniquely reconstructed from frequency ω_o and discrete-time signals $x(t_{2n})$ and $x(t_{2n+1})$ defined by equations (3.6) and (3.7), respectively, corresponding to the second-order heteromerous sampling with average sampling period $T = 2\pi/\Omega$.

Kohlenberg proved that

$$x_c(t) = \sum_{n=-\infty}^{\infty} [x(t_{2n})f_s(t - t_{2n}) + x(t_{2n+1})f_s(t_{2n+1} - t)] \ . \quad (3.28)$$

The function $f_s(t)$ is given by

$$f_s(t) = \frac{\cos\left[\frac{\pi m T_1}{T} - (\omega_o + \frac{\Omega}{2})t\right] - \cos\left[\frac{\pi m T_1}{T} - ((2m-1)\frac{\Omega}{2} - \omega_o)t\right]}{\frac{\pi \sin(\pi m T_1/T)}{T}t}$$
$$+ \frac{\cos\left[\frac{\pi(2m-1)T_1}{2T} - ((2m-1)\frac{\Omega}{2} - \omega_o)t\right] - \cos\left[\frac{\pi(2m-1)T_1}{2T} - \omega_o t\right]}{\frac{\pi \sin[\pi(2m-1)T_1/2T]}{T}t} \quad (3.29)$$

in which m is the greatest number such that $2m$ is an integer and $(m-1)\Omega/2 \leq \omega_o$. Moreover $T_1 = T$ can be chosen only when m is an integer and

$$\omega_o = (m-1)\Omega/2 \ . \quad (3.30)$$

Kohlenberg bandlimited sampling theory was formalized and generalized by Vaughan, Scott and White (1991) and Coulson (1995). New impulses to the sampling theory gave recent achievements in the wavelet transformation theory (Chui, 1992; Walter, 1992). Recently, Vaidyanathan and Phoong (1995) showed that heteromerous sampling of appropriate order can be used for stable reconstruction of non-bandlimited signals.

It is well known (Crochiere and Rabiner, 1983; Dąbrowski, 1988b) that formula (3.26) can be generalized to bandlimited signals with spectra $\omega_1 \leq |\omega| \leq \omega_2$ restricted to the so-called integer bands with respect to frequency Ω defined by

$$\omega_1 = (m-1)\Omega/2 \ , \quad \omega_2 = m\Omega/2 \tag{3.31}$$

where m is any integer greater than zero. Such signals are referred to in this book as the **integer-band signals** with respect to frequency Ω.

The following proposition follows directly from the Kohlenberg theorem: the continuous-time signal $x_c(t)$ which is an integer-band signal with respect to frequency Ω can be uniquely reconstructed from the integer parameter m, defining the band position according to expression (3.31), and from the discrete-time signal $x(t_n)$ obtained by uniform sampling with period $T = 2\pi/\Omega$.

To prove this proposition, it is enough to notice that from the fact that m is an integer it follows that equation (3.30) is fulfilled and thus $T_1 = T$ can be chosen. From expressions (3.28) and (3.29) we obtain the reconstruction formula

$$x_c(t) = \sum_{n=-\infty}^{\infty} x(t_n) f_s(t - t_n) \tag{3.32}$$

where

$$f_s(t) = \frac{T}{\pi t} \left[\sin \frac{\pi m}{T} - \sin \frac{\pi(m-1)}{T} t \right] . \tag{3.33}$$

Signals for particular applications, e.g. internal signals in transmultiplexers may have spectra divided into many bands (Fettweis, 1982). The above proposition can be generalized for a class of such signals referred in this book to as the **integer L-band signals**. Consider a set of bands $\{\mathcal{P}_{-L}, \mathcal{P}_{1-L}, \ldots, \mathcal{P}_{-1}, \mathcal{P}_1, \mathcal{P}_2, \ldots, \mathcal{P}_L\}$ given by

$$\mathcal{P}_l = [\omega_{l1}, \omega_{l2}], \ \mathcal{P}_{-l} = [-\omega_{l2}, -\omega_{l1}], \ l = 1, 2, \ldots, L \tag{3.34}$$

$$\omega_{l1} = m_l \Omega + \omega_{l-1} \text{ and } \omega_{l2} = m_l \Omega + \omega_l \tag{3.35}$$

where $m_1, m_2, \ldots m_L$ are arbitrary (positive and/or negative) integers and

$$0 = \omega_0 \leq \omega_1 \leq \cdots \leq \omega_L = \Omega/2 \ . \tag{3.36}$$

It is called the set of **integer L-bands** with respect to frequency Ω.

42   *Discrete-time systems*

The continuous-time signal $x_c(t)$ with spectrum restricted to the set $\{\mathcal{P}_{-L}, \mathcal{P}_{1-L}, \ldots, \mathcal{P}_{-1}, \mathcal{P}_1, \mathcal{P}_2, \ldots, \mathcal{P}_L\}$ of integer L-bands with respect to frequency Ω, containing no Dirac pulses δ in the band borders, i.e. in points

$$\pm \omega_{l1} \text{ and } \pm \omega_{l2} , \quad l = 1, 2, \ldots, L ,$$

is called the integer L-band signal with respect to frequency Ω.

We can now formulate the following proposition: the continuous-time signal $x_c(t)$ which is an integer L-band signal with respect to frequency Ω with bands given by (3.34), (3.35) and (3.36) can be uniquely reconstructed from integer parameters $m_1, m_2, \ldots, m_L$, from nonnegative real parameters $\omega_1, \omega_2, \ldots, \omega_{L-1}$ fulfilling conditions (3.36), and from the discrete-time signal $x(t_n)$ obtained by uniform sampling with period $T = 2\pi/\Omega$.

The proof of this proposition follows straightforwardly from the requirement of no foldover (nonrecoverable aliasing).

Notice that integer 1-band signals with respect to frequency Ω are simply integer-band signals with respect to this frequency. Indeed, for $L = 1$ we get from (3.34)–(3.36): $l = 1$, $\omega_0 = 0$, $\omega_1 = \Omega/2$, $\omega_{11} = m_1\Omega$ and $\omega_{12} = m_1\Omega + \Omega/2$. Thus, for $m_1 = 0$ we get $\mathcal{P}_1 = [0, \Omega/2]$ and the signal spectrum is restricted to $0 \leq |\omega| \leq \Omega/2$. For $m_1 = -1$ we have $\mathcal{P}_1 = [-\Omega, -\Omega/2]$ and the signal spectrum is restricted to $\Omega/2 \leq |\omega| \leq \Omega$. For $m_1 = 1$ we obtain $\mathcal{P}_1 = [\Omega, 3\Omega/2]$ and the signal spectrum is restricted to $\Omega \leq |\omega| \leq 3\Omega/2$ and so on.

3.1.5   Modified $\mathcal{Z}$ and Dirichlet transformation

According to the modified Fourier representation (3.15a) and (3.15c) of discrete-time signals, discussed in section 3.1.3, in which a nonzero offset delay t_0 defined by expression (3.5) has been taken into account, we must modify the definition of the $\mathcal{Z}$ transform

$$X(z) = \mathcal{Z}\{x(t_n)\} = z^{-t_0/T} \sum_{n=-\infty}^{\infty} x(t_n) z^{-n} = D(z)\mathcal{X}(z) \quad (3.37a)$$

where $\mathcal{X}(z)$ is the conventional $\mathcal{Z}$ transform of the discrete-time signal $x(t_n)$

$$\mathcal{X}(z) = \sum_{n=-\infty}^{\infty} x(t_n) z^{-n} \quad (3.37b)$$

and $D(z)$ is a factor corresponding to the offset delay t_0

$$D(z) = z^{-t_0/T} \ . \tag{3.37c}$$

Strictly speaking, the delay factor $D(z)$, and thus also the modified $\mathcal{Z}$ transform $\mathcal{X}(z)$, are not functions of variable z on the Gaussian plane, which is discussed below. However, we use this notation for convenience.

Function $\mathcal{X}(z)$ is also known as the **advanced** $\mathcal{Z}$ transform (Ragazzini and Franklin, 1958) because it corresponds to the classical $\mathcal{Z}$ transform of time series $x_a(nT) = x_a(t_n - t_0) = x(t_0 + nT)$ that is advanced version of signal $x(t_n)$ by time t_0. The advance of a signal is not a causal operation. Thus, we can alternatively consider signal $x_d(nT) = x_d(t_n - t_0) = x(t_0 + (n-1)T)$ delayed by time $T - t_0$, resulting in a delayed $\mathcal{Z}$ transform $z^{-1}\mathcal{Z}(z)$.

The modified inverse transformation is defined by

$$x(t_n) = \frac{1}{2\pi \mathrm{j}} \oint_C \mathcal{X}(z) z^{t_0/T + n - 1} \mathrm{d}z \tag{3.38}$$

in which the integration is performed in a positive direction along the closed line C contained in the convergence region of transformation (3.37a) and encircling the coordinate origin $z = 0$.

The complex variable z is related to the complex signal frequency $p = \sigma + \mathrm{j}\omega$ by the expression

$$z = \mathrm{e}^{pT} \ . \tag{3.39}$$

Note that although function $\mathcal{Z}(z)$ is usually analytic on the Gaussian plane, this is not true for the delay factor $D(z)$. However, the only role of this factor is to provide information about the delay t_0. Thus we certainly have to assume that

$$D(z) = \mathrm{e}^{-pt_0} \ .$$

Consequently, we also assume that pt_0 can be unambiguously determined from $D(z)$ as

$$pt_0 = -\ln D(z) \ .$$

This means that value $D(z)$, and also z by itself, must be expressed in the Riemann surface for function $\ln z$. Such a Riemann surface is constructed with infinitely many complex planes lying over each other and connected to each other along, for example, the line $re^{j\pi}$ where r is a real positive variable. Thus, we can imagine this Riemann surface as a flat infinitely wide 'screw' (or 'spiral') with infinitely many turns. This observation simply means that the argument of value $D(z)$ must be expressed in a measure extended to the whole real axis rather than in the commonly used measure restricted to an interval of length 2π, e.g. to interval $(-\pi, \pi]$. The former is referred to as the 'absolute angular measure' and the latter as the 'relative angular measure'. In other words, we assume that angles α and $\alpha + 2k\pi$, $k = \pm 1, \pm 2, \ldots$, are not equivalent, i.e. we distinguish them in the absolute measure although they are represented by the same angle in the relative measure.

The multirate systems discussed in Chapter 4 operate with various sampling rates. Assume that signals with two different sampling rates F and F_o occur, related by condition (4.2a). The respective sampling periods $T = 1/F$ and $T_o = 1/F_o$ are then related by

$$T_o = MT$$

where M is some positive integer. In this case, two complex variables can be defined: variable z defined by equation (3.39) and variable z_o given by

$$z_o = e^{pT_o} = z^M \ . \tag{3.40}$$

We say that variables z and z_o are defined with respect to sampling periods T and T_o, respectively. They should be henceforth always precisely distinguished from the context, even if subscript 'o' is omitted in the notation of the latter variable. The subscript 'o' will frequently be omitted in further considerations, unless both variables z and z_o occur together in the same expression. It should be also stressed that we will normally use that variable which is defined with respect to the actual sampling period.

Functions $\mathcal{X}(z)$ and $D(z)$ are henceforth called the **analytic part** and the **delay part** of the function $X(z)$, respectively. More-

Discrete-time signals 45

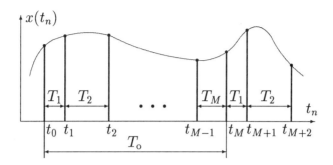

Fig. 3.3 Illustration of heteromerous sampling.

over, if we write
$$X(z) = F(z^M)$$
where M is any integer, then we will mean that the analytic part $\mathcal{X}(z)$ can be expressed as
$$\mathcal{X}(z) = \mathcal{F}(z^M)$$
where $\mathcal{F}(z)$ is an analytic function of variable z in the Gaussian plane.

The presented approach can be further generalized for nonuniformly sampled signals, resulting in the so-called **Dirichlet transformation** (Leontiev, 1976; Wojtkiewicz and Tuszyński, 1983). Consider a nonuniform but periodic, i.e. heteromerous sampling of a continuous-time signal $x_c(t)$. This process is illustrated in Fig. 3.3. The resultant sample train $x_c(t)$ has an overall period T_o composed, in general, of M nonuniform sampling periods $T_1, T_2, \ldots, T_M > 0$ so that
$$T_o = \sum_{m=1}^{M} T_m \tag{3.41}$$
and
$$t_n = t_0 + \sum_{\mu=1}^{n} T_\mu \tag{3.42}$$
for $t_0 < t_n \leq t_0 + T_o$ (i.e. for $0 < n \leq M$), while for other sampling instants corresponding multiples of T_o must be added to the right-hand side of equation (3.42).

46 *Discrete-time systems*

The Dirichlet transform, or $\mathcal{D}$ transform for short, of signal $x(t_n)$ is defined as
$$\mathcal{D}\{x(t_n)\} = \sum_n x(t_n) e^{pt_n} \ . \tag{3.43}$$
This transform is linear, unique, and for causal signals the inverse transform exists.

3.1.6 Polyphase signal decomposition

As already mentioned in section 3.1.1, polyphase decomposition of discrete-time signals is important for the description of systems with heteromerously (nonuniformly but periodically) sampled signals and for the description of multirate systems.

Let us again consider the general case of a continuous-time signal $x_c(t)$ sampled in the heteromerous manner shown in Fig. 3.3.

Notice that the resulting discrete-time signal $x(t_n)$ may be regarded as a superposition of M uniformly sampled signals
$$x_\nu(t_{\mu M+\nu}) = x(t_{\mu M+\nu}) \ ,$$
$$\mu = 0, \pm 1, \pm 2, \ldots \ , \quad \nu = 0, 1, \ldots, M-1 \ ,$$
called the **polyphase components** of the signal $x(t_n)$, as shown in Fig. 3.4.

Let $t_{\nu 0} = t_\nu$ for $\nu = 0, 1, \ldots, M-1$. Then, the $\mathcal{Z}$ transform of the νth polyphase component can be expressed as
$$X_\nu(z) = z^{-t_{\nu 0}/T_o} \sum_{\mu=-\infty}^{\infty} x_\nu(t_{\mu M+\nu}) z^{-\mu} = z^{-t_{\nu 0}/T_o} \mathcal{X}_\nu(z) \tag{3.44}$$

Consequently, the Dirichlet transform for the whole signal $x(t_n)$ is given by
$$\mathcal{D}\{x(t_n)\} = \sum_{\nu=0}^{M-1} X_\nu(z) = \sum_{\nu=0}^{M-1} e^{-pt_{\nu 0}} \mathcal{X}_\nu(z) \ . \tag{3.45}$$

Note that variable $z = e^{pT_o}$ in expressions (3.44) and (3.45) is defined with respect to period T_o.

Decomposition of the signal $x(t_n)$ into polyphase components may be considered as a collection of the following product modulations
$$x_\nu(t_{\mu M+\nu}) = q_\nu(t_n) x(t_n) \ . \tag{3.46}$$

Discrete-time signals   47

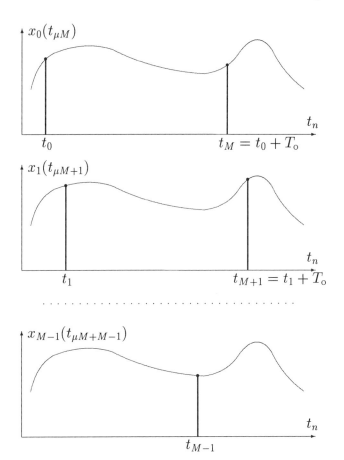

Fig. 3.4 Polyphase components.

For the νth polyphase component, the carrier signal $q_\nu(t_n)$ is composed of M-element sequences of the form

$$\tilde{q}_\nu = (q_{\nu 0}, q_{\nu 1}, \ldots, q_{\nu(M-1)}) = (\underbrace{0, 0, \ldots, 0}_{\nu \text{ zeros}}, 1, \underbrace{0, 0, \ldots, 0}_{M-\nu-1 \text{ zeros}}) \quad (3.47)$$

where

$$\ldots, q_\nu(t_{-M}) = q_\nu(t_0) = q_\nu(t_M) = \ldots = q_{\nu 0},$$

$$\ldots, q_\nu(t_{-M+1}) = q_\nu(t_1) = q_\nu(t_{M+1}) = \ldots = q_{\nu 1},$$

and so on.

48 Discrete-time systems

Including formula (3.47) in expression (3.21a) and adapting expression (3.21b), we obtain

$$Q_{\nu m} = \frac{1}{M}\sum_{n=0}^{M-1} q_\nu(t_n) W_M^{mn} = \frac{1}{M} W_M^{m\nu}, \qquad (3.48a)$$
$$m = 0, 1, \ldots, M-1$$

and

$$q_\nu(t_n) = \sum_{m=0}^{M-1} \frac{1}{M} W_M^{m\nu} W_M^{-mn} = \frac{1}{M}\sum_{m=0}^{M-1} W_M^{m(\nu-n)}. \qquad (3.48b)$$

Substituting this result into equation (3.46) and using the following notation

$$x^{(m)}(t_n) = W_M^{-mn} x(t_n) \qquad (3.49)$$

for the complex modulated signal $W_M^{-mn} x(t_n)$, we get

$$x_\nu(t_{\mu M+\nu}) = \frac{1}{M}\sum_{m=0}^{M-1} W_M^{m\nu} x^{(m)}(t_n). \qquad (3.50)$$

Further considerations in this section are restricted to a particular case, namely to the uniform sampling, which is important for the description of multirate systems. In this case, we can write

$$T_1 = T_2 = \ldots = T_M = T = T_\mathrm{o}/M$$

and

$$t_{\mu M+\nu} = t_{\nu 0} + \mu T_\mathrm{o} = t_0 + \nu T + \mu T_\mathrm{o}.$$

Consider the $\mathcal{Z}$ transform (3.44) of the νth polyphase component. On the basis of equality (3.50), this transform can be expressed by the $\mathcal{Z}$ transforms of complex modulated signals $x^{(m)}(t_n)$, $m = 0, 1, \ldots, M-1$, defined in equation (3.49) as follows:

$$X_\nu(z^M) = \frac{1}{M}\sum_{m=0}^{M-1} W_M^{m\nu} X^{(m)}(z) \qquad (3.51a)$$

where z variable is defined with respect to sampling period $T = T_\mathrm{o}/M$. For conventional $\mathcal{Z}$ transforms, this expression can be modified as follows

$$z^{-\nu}\mathcal{X}_\nu(z^M) = \frac{1}{M}\sum_{m=0}^{M-1} W_M^{m\nu} \mathcal{X}^{(m)}(z). \qquad (3.51b)$$

The $\mathcal{Z}$ transform of the complex modulated signal $x^{(m)}(t_n)$ can, in turn, be expressed by the $\mathcal{Z}$ transform of the initial discrete-time signal $x(t_n)$

$$X^{(m)}(z) = W_M^{mt_o/T} X(zW_M^m) \qquad (3.52a)$$

and similarly for conventional $\mathcal{Z}$ transforms

$$\mathcal{X}^{(m)}(z) = \mathcal{X}(zW_M^m) \ . \qquad (3.52b)$$

Finally, the $\mathcal{Z}$ transform of the initial discrete-time signal $x(t_n)$ can be written as a sum of the $\mathcal{Z}$ transforms of all polyphase components, i.e.

$$X(z) = \sum_{\nu=0}^{M-1} X_\nu(z^M) \qquad (3.53a)$$

and similarly for conventional $\mathcal{Z}$ transforms

$$\mathcal{X}(z) = \sum_{\nu=0}^{M-1} z^{-\nu} \mathcal{X}_\nu(z^M) \ . \qquad (3.53b)$$

3.2 DISCRETE-TIME PSEUDOPASSIVE SYSTEMS

3.2.1 Wave discrete-time systems

A discrete-time system can be fully described by a signal-flow graph (SFG) determining its operation (Oppenheim and Schafer, 1975). The system operates cyclically. The rate $F = 1/T$ is referred to as the **operation rate** and T is called the **operation period**. Using the terminology suggested by Fettweis (1973, 1986), we will call the system **full-synchronic** if the total delay in each branch of its SFG is an integer multiple of the operation period T. Otherwise, the system is referred to as **half-synchronic**, provided that its SFG is realizable.

A proper SFG, i.e. compact and containing no branches with zero transmittance, is realizable if the total delay in each cycle* is an integer multiple of the operation period T[†].

---

*A cycle is a directed loop, i.e. one that is composed of consistently directed branches.

[†]In the theory of digital systems, an additional statement is necessary: a realizable SFG must contain no delay-free loops. This assumption is, however, irrelevant for most of SC circuits.

50   *Discrete-time systems*

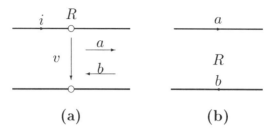

Fig. 3.5 Correspondence between a reference circuit and a wave discrete-time system: (a) reference circuit port, (b) wave SFG port.

Continuous-time systems (referred to as **reference circuits**) can be transformed into realizable discrete-time SFGs (or more precisely to digital SFGs, i.e. those containing no delay-free loops) by the bilinear transformation

$$\psi = \frac{z-1}{z+1} = \tanh\left(\frac{pT}{2}\right) \tag{3.54}$$

where z is the $\mathcal{Z}$ transformation variable (cf. equation (3.37a)), p is a complex frequency in the resulting discrete-time system and ψ is a complex frequency in the reference circuit. However, for avoidance of delay-free loops, waves should be used, instead of voltage v and current i, as primary (i.e. transformation) quantities (Fig. 3.5). We should namely use the incident wave

$$a = v + Ri \tag{3.55a}$$

and the reflected wave

$$b = v - Ri \ , \tag{3.55b}$$

where by v, i and $R > 0$ the voltage, current and the characteristic constant (i.e. wave resistance) are denoted, respectively, for a chosen port of the reference circuit (Fettweis, 1971). The transformed discrete-time system referred to as the **wave digital system** preserves all the excellent properties its reference circuit might possess, such as robust (structural) stability and low sensitivity to element variations (Fettweis, 1974). For such interesting properties, the reference circuit should be a properly designed lossless system (e.g. a classical doubly resistively-terminated ladder).

For SC circuits, the absence of delay-free loops in their SFGs is of secondary importance and therefore voltages and currents can also be successfully used as transformation quantities (Lee et al., 1981; Dąbrowski, 1982b, 1985). The use of wave quantities is, however, advantageous also in this case. Moreover, the additional degree of freedom compared to digital realization, consisting in only a formal fulfilment of equations (3.55a) and (3.55b), i.e. without the causality of wave a with respect to wave b, necessary in a digital case*, leads to entirely new class of SC networks, namely to the so-called **voltage inverter switch SC circuits** (or VIS-SC circuits for short), which possess particularly interesting properties (Fettweis, 1979a, b). One should stress that there also exist wave-SC circuits, which have the same signal-flow graphs (SFGs) as WDFs (Kleine et al., 1981; Mavor et al., 1981). They are, however, less interesting than VIS-SC circuits for they are more complex in structure and number of elements (they have e.g. more op amps).

3.2.2 Discrete-time reference circuit

In the original theory of wave digital systems, a continuous-time reference circuit concept was used (Fettweis, 1971). Therefore, the correspondence (3.54) between the reference circuit and the resulting discrete-time system was defined in the frequency domain. Such an approach is satisfactory for linear and single-rate (i.e. shift-invariant or stationary) systems only. For multirate systems, the time-domain correspondence would be much more convenient. That is why we introduce in this section a discrete-time reference circuit concept (Dąbrowski and Fettweis, 1987).

A continuous-time circuit is a dynamic system described by a set of differential algebraic equations. These can be discretized in time, e.g. by means of the trapezoidal rule (Gear, 1971). Such a discretization process is general, but if the trapezoidal rule with uniform sampling is chosen, then in a linear and shift-invariant case it will be equivalent to relation (3.54).

---

*Causality in this sense means that the sample b_n of the reflected wave is delayed with respect to the sample a_n of the incident wave.

52 Discrete-time systems

We shall henceforth assume that the continuous-time reference circuit is discretized by the trapezoidal rule with heteromerous sampling in general. The resulting discrete-time system is referred to as the **discrete-time reference circuit**. The discrete-time voltage and current are then related to the discrete-time waves by the following equations

$$v(t_n) = [a(t_n) + b(t_n)]/2 \qquad (3.56a)$$

and

$$i(t_n) = [a(t_n) - b(t_n)]/2 \; . \qquad (3.56b)$$

For instance, for a discrete resistor R, inductor R and capacitor R^{-1}, we have

$$\begin{aligned}
v(t_n) &= Ri(t_n) & (3.57a) \\
v(t_{n+1}) + v(t_n) &= R[i(t_{n+1}) - i(t_n)] & (3.57b) \\
i(t_{n+1}) + i(t_n) &= R^{-1}[v(t_{n+1}) - v(t_n)] \;, & (3.57c)
\end{aligned}$$

respectively. Conversely, solving these difference relations for uniform sampling under steady-state conditions, one easily obtains from expressions (3.37a) and (3.54) the conventional equations

$$\begin{aligned}
V &= RI & (3.58a) \\
V &= \psi RI & (3.58b) \\
I &= \psi R^{-1} V \;, & (3.58c)
\end{aligned}$$

respectively.

3.2.3 Pseudopower and pseudoenergy

In the literature, e.g. Oppenheim and Schafer (1975), 'energy' and 'power' are often used to describe quantities in the discrete-time signal domain. In this book, however, the notions **pseudoenergy** and **pseudopower** are consequently used to distinguish these formally defined discrete-time domain quantities from the physical energy and power corresponding to physical representations of discrete-time signals.

Discrete-time pseudopassive systems   53

The instantaneous signal peudopower of a real-valued signal $x(t_n)$ is defined as
$$p_{sx}(t_n) = x^2(t_n) \ . \tag{3.59}$$
Moreover
$$\varepsilon_{sx}(t_n) = \sum_{\nu=-\infty}^{n} p_{sx}(t_\nu) \tag{3.60}$$
and
$$\epsilon_{sx} = \lim_{n \to \infty} \varepsilon_{sx}(t_n) \tag{3.61}$$
are quantities referred to as the partial signal $x(t_n)$ pseudoenergy up to instant t_n and the total signal pseudoenergy, respectively. The signal $x(t_n)$ is called a finite pseudoenergy signal if $\epsilon_{sx} < \infty$. Finite pseudoenergy signals fulfil the following condition
$$\bigwedge_{t_n} \varepsilon_{sx}(t_n) < \infty \ . \tag{3.62}$$
The reverse statement is, however, not true.

Important also is the class of causal signals defined as
$$\{x(t_n) : \bigvee_{n_o} \bigwedge_{n<n_o} x(t_n) = 0\} \tag{3.63}$$
where n_o is a certain integer.

The signal pseudoenergy ϵ_{sx} can also be expressed with the signal spectrum $X(e^{j\omega T})$ (cf. (3.15a)) by the following Parseval equation (Oppenheim and Schafer, 1975)
$$\begin{aligned}\epsilon_{sx} = \sum_{n=-\infty}^{\infty} x^2(t_n) &= \frac{1}{\Omega} \int_{-\Omega/2}^{\Omega/2} \left|X\left(e^{j\omega T}\right)\right|^2 d\omega \\ &= \frac{2}{\Omega} \int_{0}^{\Omega/2} \left|X\left(e^{j\omega T}\right)\right|^2 d\omega \ . \end{aligned} \tag{3.64a}$$

For a finite M-element signal $q(t_n)$, $n = 0, 1, \ldots, M-1$, with DFT representations (3.19a) and (3.21a), the Parseval equation has the following finite form
$$\begin{aligned}\epsilon_{sq} = \sum_{n=0}^{M-1} q^2(t_n) &= M \sum_{m=0}^{M-1} |Q(m\Omega_o)|^2 \\ &= M \sum_{m=0}^{M-1} |Q_m|^2 \ . \end{aligned} \tag{3.64b}$$

54 Discrete-time systems

In addition to the above quantities describing isolated signals, we define pseudopower and pseudoenergy in a discrete-time reference circuit and in the resulting (wave) discrete-time system.

In a chosen port of a discrete-time reference circuit, the instantaneous reference pseudopower $p_r(t_n)$, partial reference pseudoenergy $\varepsilon_r(t_n)$ and total reference pseudoenergy ϵ_r are defined as

$$p_r(t_n) = v(t_n)i(t_n) , \tag{3.65}$$

$$\varepsilon_r(t_n) = \sum_{\nu=-\infty}^{n} p_r(t_\nu) \tag{3.66}$$

and

$$\epsilon_r = \lim_{n\to\infty} \varepsilon_r(t_n) , \tag{3.67}$$

respectively.

Furthermore, the instantaneous wave pseudopower $p_w(t_n)$, partial wave pseudoenergy $\varepsilon_w(t_n)$ and total wave pseudoenergy ϵ_w in a chosen port of a wave discrete-time system are defined as

$$p_w(t_n) = R^{-1}[a^2(t_n) - b^2(t_n)] , \tag{3.68}$$

$$\varepsilon_w(t_n) = \sum_{\nu=-\infty}^{n} p_w(t_\nu) \tag{3.69}$$

and

$$\epsilon_w = \lim_{n\to\infty} \varepsilon_w(t_n) , \tag{3.70}$$

respectively.

From expressions (3.56a), (3.56b), (3.59), (3.65) and (3.68) we immediately get the following relations

$$p_w(t_n) = 4p_r(t_n) = R^{-1}[p_{sa}(t_n) - p_{sb}(t_n)] \tag{3.71a}$$

$$\varepsilon_w(t_n) = 4\varepsilon_r(t_n) = R^{-1}[\varepsilon_{sa}(t_n) - \varepsilon_{sb}(t_n)] \tag{3.71b}$$

$$\epsilon_w = 4\epsilon_r = R^{-1}[\epsilon_{sa} - \epsilon_{sb}] . \tag{3.71c}$$

3.2.4 Pseudopassivity and pseudolosslessness

Notions of pseudopassivity and pseudolosslessness were introduced into the theory of discrete-time systems in the context of voltage wave digital filters (VWDFs) (Fettweis, 1972). This concept was then generalized and also adopted for:

- digital ladder filters (Bruton and Vaughan-Pope, 1976)

- orthogonal filters, i.e. power wave digital filters (PWDFs) (Deprettere and Dewilde, 1980; Piekarski and Zarzycki, 1986)

- linear bounded real filters (LBRFs) (Vaidyanathan, 1985; Vaidyanathan and Mitra, 1985)

- digital lattice filters (Gray and Markel, 1973; Gray, 1980; Vaidyanathan, 1986)

- general paraunitary systems (Vaidyanathan, 1993).

As a matter of fact, the pseudopassivity and pseudolosslessness concept can be adopted to many other methods and procedures of signal processing, including adaptive algorithms such as recursive least squares (RLS) (McWhirter and Proudler, 1992), multirate systems (Dąbrowski and Fettweis, 1987), FFT algorithms (e.g. Cooley and Tukey, 1965), etc., based on lossless operations: delays, wave adaptors, Givens rotations (Givens, 1958), complex rotations.

The total pesudoenergy stored in delays of a pseudopassive system plays the role of the Liapunov function and is useful in a qualitative system stability analysis. Properly designed pseudolossless systems possess many advantages. Among the most important are the following:

- low passband sensitivity to parameter changes

- low level of roundoff noise

- robustness against parasitic oscillations

- great stability margin under looped conditions, e.g. in a communication link (Fettweis and Meerkötter, 1977; Dąbrowski, 1986b)

56 Discrete-time systems

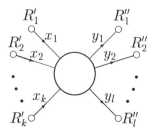

Fig. 3.6 Generalized wave discrete-time system.

- possibility for magnitude distortion free or even perfect reconstruction of signals split into into subbands (cf. section 4.5.2).

Consider the generalized wave discrete-time system shown in Fig. 3.6. The generalization in comparison to the structure in Fig. 3.5b consists in assuming individual characteristic constants for each signal, not for a port (i.e. a pair of signals containing one input and one output signal). The instantaneous wave pseudopower absorbed in instant t_n by the system in Fig. 3.6 is defined as

$$\begin{aligned} p(t_n) &= \sum_{\kappa=1}^{k} R'^{-1}_\kappa x^2_\kappa(t_n) - \sum_{\lambda=1}^{l} R''^{-1}_\lambda y^2_\lambda(t_n) \\ &= \mathbf{x}^T(t_n)\mathbf{G}'\mathbf{x}(t_n) - \mathbf{y}^T(t_n)\mathbf{G}''\mathbf{y}(t_n) \end{aligned} \qquad (3.72)$$

where

$$\mathbf{G}' = \operatorname{diag}\left(R'^{-1}_1, R'^{-1}_2, \ldots, R'^{-1}_k\right) \qquad (3.73a)$$
$$\mathbf{G}'' = \operatorname{diag}\left(R''^{-1}_1, R''^{-1}_2, \ldots, R''^{-1}_l\right) \qquad (3.73b)$$

and by $\mathbf{x}^T(t_n)$, $\mathbf{y}^T(t_n)$ vectors transposed to vectors

$$\mathbf{x}(t_n) = \begin{bmatrix} x_1(t_n) \\ x_2(t_n) \\ \vdots \\ x_k(t_n) \end{bmatrix} \quad \text{and} \quad \mathbf{y}(t_n) = \begin{bmatrix} y_1(t_n) \\ y_2(t_n) \\ \vdots \\ y_l(t_n) \end{bmatrix}.$$

Consider a steady state in this system assuming that complex excitations

$$x_\kappa(t_n) = X_\kappa e^{pt_n}, \quad \kappa = 1, 2, \ldots, k \qquad (3.74)$$

are applied to it, where X_κ and $p = \sigma + j\omega$ are complex constants. Then, all signals in the system, including output signals, are also of the form

$$y_\lambda(t_n) = Y_\lambda e^{pt_n} \;\;, \lambda = 1, 2, \ldots, l \tag{3.75}$$

We shall henceforth assume that r.m.s. values are used throughout. This means that the values of samples of the discrete-time signals actually occuring are $2^{1/2}$ times the real part of the complex signals (3.74) and (3.75).

From equation (3.72), the average absorbed wave pseudopower P can be computed as

$$\begin{aligned} P &= \sum_{\kappa=1}^{k} R'^{-1}_\kappa |X_\kappa|^2 - \sum_{\lambda=1}^{l} R''^{-1}_\lambda |Y_\lambda|^2 \\ &= \mathbf{X}^* \mathbf{G}' \mathbf{X} - \mathbf{Y}^* \mathbf{G}'' \mathbf{Y} \end{aligned} \tag{3.76}$$

where $\mathbf{X}^*$ and $\mathbf{Y}^*$ are vectors conjugate to transposed vectors

$$\mathbf{X}^T = \begin{bmatrix} X_1 & X_2 & \cdots & X_k \end{bmatrix} \text{ and } \mathbf{Y}^T = \begin{bmatrix} Y_1 & Y_2 & \cdots & Y_l \end{bmatrix}.$$

The system in Fig. 3.3 will be called **externally pseudopassive** if the following condition is fulfilled

$$\bigwedge_{\mathbf{X}} \operatorname{Re} p \geq 0 \Rightarrow P \geq 0 \;. \tag{3.77}$$

If additionally the condition

$$\bigwedge_{\mathbf{X}} p = j\omega \Rightarrow P = 0 \tag{3.78}$$

holds, then the system is referred to as **externally pseudolossless**.

Note, for example, from the Parseval equation (3.64b), that the system computing DFT defined by expression (3.21a) (or a slightly more general system realizing equation (3.19a)) is a pseudolossless system*. This is immediately clear from expressions (3.72) and (3.76) with the following choice for variables in equation (3.21a): $x_\kappa(t_n) = q_{\kappa-1}$, $y_\lambda(t_n) = Q_{\lambda-1}$ and $R'_\kappa = 1$, $R''_\lambda = M^{-1}$ for $\kappa, \lambda = 1, 2, \ldots, M$.

---

*In fact, it is even a static, pseudoenergy neutral system.

58   *Discrete-time systems*

The system in Fig. 3.6 is composed of two kinds of building-blocks: dynamic elements (delays) and static elements (performing arithmetic relations). Delays are obviously pseudolossless operations, thus the system will be referred to as **internally pseudopassive (pseudolossless)** if all its static elements are pseudopassive (pseudolossless). Voltage inverter switches (VISs), adaptors (i.e. static blocks in wave discrete-time systems), Givens rotators, complex rotators and DFTs/IDFTs can serve as examples of pseudolossless static building-blocks.

Internal pseudopassivity (pseudolosslessness) is stronger than external pseudopassivity (pseudolosslessness), i.e. internally pseudopassive (pseudolossless) systems are always externally pseudopassive (pseudolossless) but not vice versa. All the marvellous advantages listed above, such as low passband sensitivity, low level roundoff noise, robustness against parasitic oscillations and a great stability margin under looped conditions, are valid only for internally pseudolossless systems.

A typical pseudolossless wave discrete-time arrangement is shown in Fig. 3.7. It has two inputs (incident waves – signals a_1 and a_2) and two outputs (reflected waves – signals b_1 and b_2). The inputs are formed by the incident waves corresponding to two ports of a reference circuit* and the outputs are formed by the respective reflected waves. Such a system realizes a filter bank (cf. section 4.5) of four single-input/single-output discrete-time filters with transfer functions denoted in Fig. 3.7 as S_{11}, S_{12}, S_{21} and S_{22}

$$\begin{bmatrix} B_1 \\ B_2 \end{bmatrix} = \mathbf{S} \begin{bmatrix} A_1 \\ A_2 \end{bmatrix} = \begin{bmatrix} S_{11} & S_{12} \\ S_{21} & S_{22} \end{bmatrix} \begin{bmatrix} A_1 \\ A_2 \end{bmatrix} \quad (3.79)$$

where A_1, A_2 and B_1, B_2 are $\mathcal{Z}$ transforms of the respective signals. The static block is a pseudolossless n-port composed in a digital case of adaptors and sign inverters (multipliers by -1) (Fettweis, 1986).

Assuming that signals a_1, a_2 and b_1, b_2 describe power waves, i.e. choosing constant $R = 1$ in expression (3.68), thus consequently choosing $R'_\kappa = R''_\lambda = 1$, $X_\kappa = A_\kappa$ and $Y_\lambda = B_\lambda$ ($\kappa, \lambda = 1, 2$) in

---
*We assume that the reference circuit is a two-port circuit.

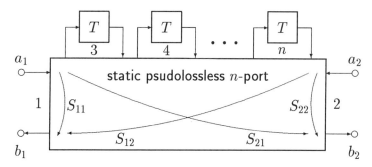

Fig. 3.7 Wave discrete-time pseudolossless system.

equation (3.76), we can write for a pseudolossless system for $p = j\omega$

$$\mathbf{S}^*\mathbf{S} = \mathbf{1} \tag{3.80}$$

where $\mathbf{S}^*$ is obtained from $\mathbf{S}$ by transposing it and replacing its entries by their complex conjugate. A consequence of this equation is the so-called Feldtkeller relations

$$|S_{11}|^2 + |S_{21}|^2 = 1 \text{ and } |S_{22}|^2 + |S_{12}|^2 = 1 \ . \tag{3.81}$$

Moreover, it is known (Belevitch, 1968) that

$$|S_{12}| = |S_{21}| \ . \tag{3.82}$$

Hence, these transfer functions form four pairs of complementary filters, namely (S_{11}, S_{21}), (S_{12}, S_{22}), (S_{11}, S_{12}) and (S_{21}, S_{22}). In a reciprocal and symmetrical case, these filters are not only pair-wise complementary but also pair-wise identical, i.e. $S_{11} = S_{22}$ and $S_{12} = S_{21}$. Two of them are, for instance, identical low-pass filters while the other two are identical complementary high-pass filters. Thus, wave digital arrangements inherently realize **branching filter banks***. Design of branching filter banks as wave discrete-time systems seems to be a very good way not only because of the fact

---

*A so-called branching (or directional) filter bank is a pair of complementary filters H and $\bar{H}$. In the most typical case, one of them is a low-pass filter, while the other is a complementary high-pass filter (Fig. 4.20). In the case of a symmetrical band partition, it is often reasonable to connect branching filtering with sampling rate alteration by a factor of 2; the sampling rate is reduced by a factor of 2 when splitting the signal into two bands and is increased by a factor of 2 when composing the signal from two bands (cf. section 4.5.1).

that both bank filters are realized in one structure only but also due to many other substantial advantages of wave discrete-time systems. Among the most important are:

- small passband sensitivity to coefficients
- large resistivity to parasitic oscillations
- possibility of recovery of effective pseudoenergy, normally partly lost in a process of interpolation or decimation (cf. section 8.3).

Branching filter banks (S_{11}, S_{21}) and (S_{12}, S_{22}) have one input signal (a_1 or a_2, respectively) and two output signals b_1 and b_2. They are examples of so-called **analysis filter banks** (cf. section 4.5). Bank (S_{11}, S_{21}) will be realized if we put $a_2 = 0$. Similarly, if we assume $a_1 = 0$, bank (S_{12}, S_{22}) will be obtained.

On the other hand, branching filter banks (S_{11}, S_{12}) and (S_{21}, S_{22}) have two input signals a_1, a_2 and one output signal b_1 or b_2, respectively. They are known as **synthesis filter banks** (cf. section 4.5). Signal b_2 is not used in bank (S_{11}, S_{12}). Similarly, in bank (S_{21}, S_{22}), signal b_1 is not used.

In the simplest common case, however, the system in Fig. 3.7 represents a single-input/single-output filter with signal a_1 used as the input and signal b_2 used as the output. Thus, signal b_1 is not used and signal a_2 is assumed to be equal to zero.

Relation (3.80) is restricted to the unit circle in the $z = e^{pT}$ plane. For SC circuits, entries of matrix $\mathbf{S}(z)$ in expression (3.79) can usually be assumed to be rational functions of variable z and then they have real coefficients. Thus, we can analytically extend relation (3.80) to the whole z plane, resulting in

$$\mathbf{S}^T(z^{-1})\mathbf{S}(z) = 1 \qquad (3.83)$$

where $\mathbf{S}^T$ is a transposition of matrix $\mathbf{S}$. Matrix $\mathbf{S}(z)$ fulfilling condition (3.83) is called the **paraunitary matrix** (Vaidyanathan, 1993) and the respective system is said to be the **paraunitary system**. Systems which are stable and paraunitary are pseudolossless. Usually, considerations are restricted to stable systems and therefore the terms 'pseudolosslessness' and 'paraunitariness' are practically synonymous.

4

Multirate systems

As already stated in section 1.3, the notion of multirate signal processing, usually used in a narrow sense restricted to the problem of sampling rate alteration, is generalized in this book and used for the process of product modulation cascaded with filtering. Sampling rate alteration and nonuniform (heteromerous) sampling can then be considered as particular cases of this general process.

To introduce this generalized multirate system concept, it is essential first to consider how the sampling rate can be changed, and what effect this change has on the signal spectrum. This topic is presented in section 4.1. Then product modulation is considered and its relation to sampling rate alteration and to nonuniform (heteromerous) sampling are shown, followed by the concept of the general multirate building-block. Next, the topic of pseudoenergy losses of multirate-processed signals is covered and possibilities for the recovery of the lossed pseudoenergy are presented. Finally, filter banks are discussed. First, the most important principles of two-channel filter banks are treated. Then more general, modulated M-channnel filter banks are considered. Finally, polyphase filter banks providing the highest efficiency are presented.

4.1 SAMPLING RATE ALTERATION

Consider two discrete-time signals

$$x(t_n) = x(t_0 + nT) \ , \ 0 \le t_0 < T \ , \ n = 0, \pm 1, \pm 2, \ldots \quad (4.1a)$$

and

$$x_o(\tau_\nu) = x_o(\tau_0 + \nu T_o) \ , \ 0 \le \tau_0 < T_o \ , \ \nu = 0, \pm 1, \pm 2, \ldots \quad (4.1b)$$

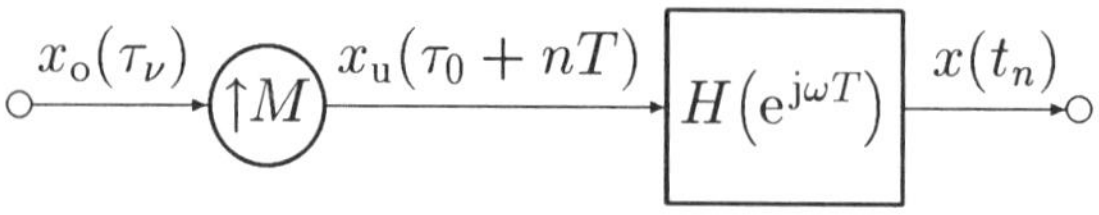

Fig. 4.1 Interpolator scheme.

obtained by uniform sampling of the same continuous-time signal $x(t)$ with two sampling periods T and T_o. Sampling rate alteration consists in converting one of these signals into the second one.

Assume that
$$T_o = MT \ , \ F = MF_o \qquad (4.2a)$$
with M being an integer called the multirate period and
$$F = 1/T \ , \ F_o = 1/T_o \qquad (4.2b)$$
being the respective sampling rates. Reconstruction of signal $x(t_n)$ from signal $x_o(\tau_\nu)$ (M-fold sampling rate increase) is referred to as **interpolation** with factor M. Similarly, the formation of signal $x_o(\tau_\nu)$ from signal $x(t_n)$ (M-fold sampling rate decrease) is referred to as **decimation**[*] with factor M.

4.1.1 Interpolation

Interpolation can be realized using the system in Fig. 4.1, i.e. by means of **up-sampling** (sampling rate expansion) followed by appropriate **anti-imaging filtering**.

Up-sampling, denoted in Fig. 4.1 by $\uparrow M$, consists in placing $M-1$ equally spaced and zero-valued samples between each pair of neighbouring original samples (Fig. 4.2). The corresponding building-block is called an **up-sampler** or **sampling rate expander**. The resulting signal is given by

$$x_u(\tau_0 + nT) = \begin{cases} x_o(\tau_\nu) & \text{for } n = \nu M \text{ and } \tau_\nu = \tau_0 + \nu T_o \\ 0 & \text{for } n \neq \nu M \end{cases} \qquad (4.3)$$

From expression (3.15a) we immediately conclude that both sig-

---
[*]Note that the term 'decimation' does not mean that the factor M is a multiple of ten.

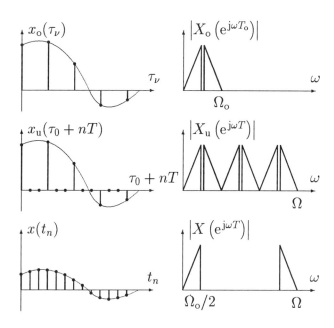

Fig. 4.2 Illustration of interpolation with factor $M = 3$; signals correspond to a lowband signal with respect to frequency Ω_o.

nals, the original signal $x_o(\tau_\nu)$ and its up-sampled version $x_u(\tau_0 + nT)$, possess the same spectrum

$$X_u\left(e^{j\omega T}\right) = X_o\left(e^{j\omega T_o}\right) \qquad (4.4a)$$

(cf. Fig. 4.2).

On the other hand, for the modified $\mathcal{Z}$ transforms of these signals we get

$$\begin{aligned} X_u(z) &= z^{-\tau_0/T} \sum_{n=-\infty}^{\infty} x_u(\tau_0 + nT) z^{-n} \\ &= (z^M)^{-\tau_0/T_o} \sum_{\nu=-\infty}^{\infty} x_o(\tau_0 + \nu T_o)(z^M)^{-\nu} \\ &= X_o(z^M) \: . \end{aligned} \qquad (4.4b)$$

A similar relation is also valid for the conventional $\mathcal{Z}$ transforms of these signals

$$\mathcal{X}_u(z) = \mathcal{X}_o(z^M) \qquad (4.4c)$$

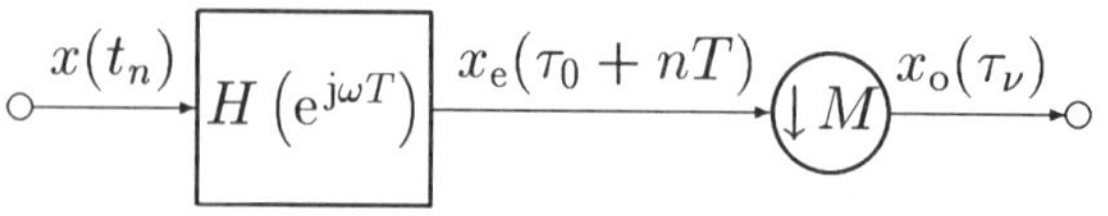

Fig. 4.3 Decimator scheme.

where z variable is defined with respect to sampling period T.

The anti-imaging filter in Fig. 4.1 operates with rate F and serves for determination of the proper output signal spectrum $X(e^{j\omega T})$ by eliminating the undesired spectral components (images) in the up-sampled signal spectrum $X_u(e^{j\omega T})$, as shown in Fig. 4.2. If the continuous-time signal $x_c(t)$ corresponding to the processed signal $x_o(\tau_\nu)$ is a lowband signal with respect to frequency Ω_o then the anti-imaging filter is a low-pass filter with ideal transfer function given by

$$H\left(e^{j\omega T}\right) = \begin{cases} \text{const} & \text{for } |\omega - k\Omega| \leq \Omega_o/2, \; k = 0, \pm 1, \pm 2, \ldots \\ 0 & \text{otherwise} \end{cases} \quad (4.5)$$

where

$$\Omega = 2\pi/T, \quad \Omega_o = 2\pi/T_o.$$

In the general case, i.e. for an integer L-band signal $x_c(t)$ with respect to frequency Ω_o with spectrum restricted to the set of bands $\{\mathcal{P}_{-L}, \mathcal{P}_{1-L}, \ldots, \mathcal{P}_{-1}, \mathcal{P}_1, \mathcal{P}_2, \ldots, \mathcal{P}_L\}$ defined by expressions

$$\mathcal{P}_l = [\omega_{l1}, \omega_{l2}], \; \mathcal{P}_{-l} = [-\omega_{l2}, -\omega_{l1}], \; l = 1, 2, \ldots, L \quad (4.6)$$

$$\omega_{l1} = m_l \Omega_o + \omega_{l-1} \text{ and } \omega_{l2} = m_l \Omega_o + \omega_l \quad (4.7)$$

and

$$0 = \omega_0 \leq \omega_1 \leq \cdots \leq \omega_L = \Omega_o/2, \quad (4.8)$$

the desired transfer function $H(e^{j\omega T})$ is more complicated, given by

$$H\left(e^{j\omega T}\right) = \begin{cases} \text{const} & \text{for } \omega \in \bigcup_{k=-\infty}^{\infty} \bigcup_{l=1}^{L} [\omega_{l1} + k\Omega, \omega_{l2} + k\Omega] \\ & \cup [-\omega_{l2} + k\Omega, -\omega_{l1} + k\Omega] \\ 0 & \text{otherwise} \end{cases}$$

$$(4.9)$$

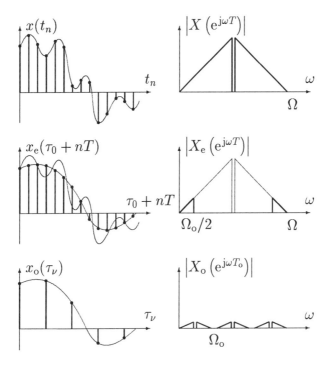

Fig. 4.4 Illustration of decimation with factor $M = 3$; signals correspond to a lowband signal with respect to frequency Ω_o.

4.1.2 Decimation

Decimation can be realized by means of the system shown in Fig. 4.3, which is called a **decimator**. The input signal $x(t_n)$ is first filtered with the **anti-aliasing filter** $H(e^{j\omega T})$ in order to obtain the effective signal $x_e(\tau_0 + nT)$, and then down-sampled simply by omitting the unnecessary samples, thus

$$x_o(\tau_\nu) = x_e(\tau_\nu) = x_e(\tau_0 + \nu T_o) \ .$$

The latter operation is performed by the so-called **down-sampler** or in other words a **sampling rate compressor**, denoted in Fig. 4.3 by $\downarrow M$.

Taking into account that signals $x_e(\tau_0 + nT)$ and $x_o(\tau_\nu)$ correspond to the same continuous-time signal, say $x_c(t)$, we conclude from equation (3.17a) that

$$X_e\left(e^{j\omega T}\right) = \frac{1}{T} \sum_{n=-\infty}^{\infty} X_c(j\omega + jn\Omega)e^{jn\Omega\tau_0} \qquad (4.10a)$$

and

$$X_o\left(e^{j\omega T_o}\right) = \frac{1}{T_o} \sum_{\nu=-\infty}^{\infty} X_c(j\omega + j\nu\Omega_o)e^{j\nu\Omega_o\tau_0} , \qquad (4.10b)$$

where

$$\Omega = 2\pi/T , \quad \Omega_o = 2\pi/T_o , \quad T_o = MT .$$

Notice that equation (4.10a) can be transformed by obvious manipulations on both sides to the following M equivalent forms

$$\frac{1}{M} X_e\left(e^{j\omega T}\right) = \frac{1}{T_o} \sum_{n=-\infty}^{\infty} X_c(j\omega + jnM\Omega_o)e^{jnM\Omega_o\tau_0} ,$$

$$\frac{1}{M} e^{-j\Omega_o\tau_0} X_e\left(e^{j(\omega - \Omega_o)T}\right) =$$
$$\frac{1}{T_o} \sum_{n=-\infty}^{\infty} X_c(j\omega + j(nM - 1)\Omega_o)e^{j(nM-1)\Omega_o\tau_0} ,$$

$$\vdots$$

$$\frac{1}{M} e^{-j(M-1)\Omega_o\tau_0} X_e\left(e^{j(\omega - (M-1)\Omega_o)T}\right) =$$
$$\frac{1}{T_o} \sum_{n=-\infty}^{\infty} X_c(j\omega + j(nM - M + 1)\Omega_o)e^{j(nM-M+1)\Omega_o\tau_0} .$$

Summing these equations together, we obtain the right-hand side of equation (4.10b). Thus the considered spectra are related as follows:

$$X_o\left(e^{j\omega T_o}\right) = \frac{1}{M} \sum_{m=0}^{M-1} e^{-jm\Omega_o\tau_0} X_e\left(e^{j(\omega - m\Omega_o)T}\right) . \qquad (4.11a)$$

With the notation of (3.22)

$$W_M = e^{-j2\pi/M} = e^{-j\Omega_o T}$$

we can finally write

$$X_o\left(e^{j\omega T_o}\right) = \frac{1}{M} \sum_{m=0}^{M-1} W_M^{m\tau_0/T} X_e\left(e^{j\omega T} W_M^m\right) . \qquad (4.11b)$$

Hence, the modified $\mathcal{Z}$ transforms are similarly related

$$\mathcal{X}_o\left(z^M\right) = \frac{1}{M}\sum_{m=0}^{M-1} W_M^{m\tau_0/T} \mathcal{X}_e\left(zW_M^m\right) . \qquad (4.12a)$$

Using now expression (3.37a), we can in turn find the following relation for the conventional $\mathcal{Z}$ transforms

$$\mathcal{X}_o\left(z^M\right) = \frac{1}{M}\sum_{m=0}^{M-1} \mathcal{X}_e\left(zW_M^m\right) . \qquad (4.12b)$$

Note that the z variable in expressions (4.12a) and (4.12b) is defined with respect to sampling period T.

An anti-aliasing filter, described by the transfer function $H(\mathrm{e}^{\mathrm{j}\omega T})$, operates with a rate F. It determines the effective signal spectrum $X_\mathrm{e}(\mathrm{e}^{\mathrm{j}\omega T})$ by reducing the spectrum $X(\mathrm{e}^{\mathrm{j}\omega T})$ of the input signal $x(t_n)$ in such a way that after the down-sampling no overlapping (non-recoverable aliasing) occurs (Fig. 4.4). It is obvious that if the input signal is lowband with respect to frequency Ω_o, then the ideal transfer function $H(\mathrm{e}^{\mathrm{j}\omega T})$ is determined by expression (4.5). In the general case, i.e. for an integer L-band signal with respect to a frequency Ω_o corresponding to the signal $x_\mathrm{e}(\tau_0+nT)$, the ideal transfer function $H(\mathrm{e}^{\mathrm{j}\omega T})$ is given by expression (4.9), similarly as that for interpolation.

4.2 PRODUCT MODULATION

Product modulation of a carrier signal $q(t_n)$ by an input signal $x(t_n)$ consists of computing the output signal

$$y(t_n) = q(t_n)x(t_n) , \quad t_n = t_0 + nT \quad n = 0, \pm 1, \pm 2, \ldots \qquad (4.13)$$

as illustrated in Fig. 4.5. We assume that the carrier signal $q(t_n)$ is periodic with period T_o related with the sampling period T according to equation (3.18). Thus, we can write

$$q(t_n + T_\mathrm{o}) = q\big(t_0 + (n+M)T\big) = q(t_n)$$

where M is a positive integer.

68 Multirate systems

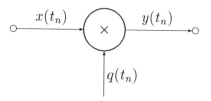

Fig. 4.5 Product modulator.

Samples within one period of the signal $q(t_n)$ form an M-element sequence, e.g. the following sequence

$$\tilde{q} = (q_0, q_1, \ldots, q_{M-1}) = (q(t_0), q(t_1), \ldots, q(t_{M-1})) \quad (4.14)$$

which is referred to as the **modulation sequence**. Its Fourier representation is given by expressions (3.21a) and (3.21b).

Consider now the problem of reconstruction of a continuous-time signal $x_c(t)$ corresponding to the input signal $x(t_n)$, on the basis of the modulated signal $y(t_n)$. From equations (3.15a), (3.19b) and (4.13) we get

$$\begin{aligned} Y\left(e^{j\omega T}\right) &= \sum_{n=-\infty}^{\infty} q(t_n)x(t_n)e^{-j\omega t_n} \\ &= \sum_{n=-\infty}^{\infty} \sum_{m=0}^{M-1} Q(m\Omega_o)e^{jm\Omega_o t_n} x(t_n)e^{-j\omega t_n} \\ &= \sum_{m=0}^{M-1} Q(m\Omega_o) \sum_{n=-\infty}^{\infty} x(t_n)e^{-j(\omega - m\Omega_o)t_n} . \end{aligned}$$

Thus, using the notation of (3.22)

$$W_M = e^{-j2\pi/M} = e^{-j\Omega_o T}$$

we conclude that

$$\begin{aligned} Y\left(e^{j\omega T}\right) &= \sum_{m=0}^{M-1} Q(m\Omega_o)X\left(e^{j(\omega - m\Omega_o)T}\right) \\ &= \sum_{m=0}^{M-1} Q(m\Omega_o)X\left(e^{j\omega T}W_M^m\right) . \quad (4.15a) \end{aligned}$$

This equation can be rewritten for $\mathcal{Z}$ transforms in the form

$$Y(z) = \sum_{m=0}^{M-1} Q(m\Omega_o) X\left(zW_M^m\right) . \quad (4.15b)$$

Consequently, using relations (3.23) and (3.37a) results in the following expression

$$\mathcal{Y}(z) = \sum_{m=0}^{M-1} Q_m \mathcal{X}(zW_M^m) . \qquad (4.15c)$$

for conventional $\mathcal{Z}$ transforms. Equations (4.15a)–(4.15c) play an important role in the analysis of multirate systems discussed in this book.

Expression (4.15a) shows that a nonrecoverable aliasing (i.e. overlapping of translated spectra $X(e^{j(\omega-m\Omega_o)T})$) can occur in $Y(e^{j\omega T})$ even if the signal $x_c(t)$ fulfils the conditions in section 3.1.4, which are sufficient for sampling with period T, i.e. if $x_c(t)$ is a lowband signal, or generally, an integer L-band signal with respect to frequency Ω. However, if stronger conditions are fulfilled, as in the case of interpolation and decimation, i.e. if the signal $x_c(t)$ is a lowband signal, or generally, an integer L-band signal with respect to frequency Ω_o, then nonrecoverable aliasing will not occur. Thus, the following proposition can be formulated: a continuous-time signal $x_c(t)$ which is an integer L-band signal with respect to frequency Ω_o (with bands given by (3.34), (3.35) and (3.36) but with Ω replaced by Ω_o) can be uniquely reconstructed from integer parameters $m_1, m_2, \ldots, m_L$, from nonnegative real parameters $\omega_1, \omega_2, \ldots, \omega_{L-1}$ fulfilling conditions (3.36) with Ω replaced by Ω_o, from a complex parameter $Q(m\Omega_o)$ defined by (3.19a) (e.g. that for $m = 0$) and from the discrete-time signal $y(t_n)$ given by (4.13) where $q(t_n)$ is a periodic carrier signal with period T_o and $x(t_n)$ is a signal resulting from sampling the signal $x_c(t)$ with period $T = 2\pi/\Omega$.

To prove this proposition it is enough to notice that

$$X\left(e^{j\omega T}\right) = Q^{-1}(0) H\left(e^{j\omega T}\right) Y\left(e^{j\omega T}\right)$$

where

$$Q(0) = Q(m\Omega_o)\bigg|_{m=0}$$

and the transfer function $H(e^{j\omega T})$ is given by expression (4.5), or generally by expression (4.9), with const $= 1$.

Now we will analyse further properties of the modulation sequence $\tilde{q}$ given by expression (4.14). The sequence pseudoenergy is

70 Multirate systems

defined as
$$\varepsilon_q = \sum_{n=0}^{M-1} q^2(t_n) \ . \qquad (4.16)$$

Thus, the sequence average pseudopower is
$$P_q = \varepsilon_q/M \ . \qquad (4.17)$$

Using the discrete-time Parseval equality (Oppenheim and Schafer, 1975), we can write
$$P_q = \frac{1}{M}\sum_{n=0}^{M-1}|q(t_n)|^2 = \sum_{m=0}^{M-1}|Q_m|^2 \qquad (4.18)$$

where Q_m, $m = 0, 1, \ldots, M-1$ is the discrete Fourier transform of the sequence $\tilde{q}$, cf. equation (3.21a).

As already stressed in section 1.3, for realizations with SC circuits, the most important sequences are modulation sequences containing only values 0 and ± 1 because in this case no multiplications are necessary. Such sequences will be referred to as **elementary sequences**. Denoting by M_1 the number of elements equal to ± 1 in an M-element elementary sequence $\tilde{q}$, equations (4.16) and (4.17) can be simplified as follows:
$$\varepsilon_q = M_1 \qquad (4.19)$$

and
$$P_q = M_1/M \ , \qquad (4.20)$$

respectively.

One of the most interesting waveforms for a carrier signal $q(t_n)$ in the product modulation (4.13) is the sinusoidal waveform. A sinusoidal carrier signal composed of M-element modulation sequences can be written in the form
$$q(t_n) = \sqrt{2}|Q|\cos(2\pi M^{-1}n + \vartheta) \ , \qquad (4.21)$$

where
$$|Q| > 0, \ -\pi < \vartheta \le \pi, \ n = 0, \pm 1, \pm 2, \ldots \ .$$

If $M > 1$, then the signal frequency f is related to the sampling rate F and the sequence length M by the following equation
$$f = F/M \ .$$

It should be stressed that, for a given sinusoidal signal $q(t_n)$, parameters $|Q|$ and ϑ in expression (4.21) are uniquely determined only if $M \geq 3$; whereas for $M = 1$ or $M = 2$ parameter ϑ can be chosen arbitrarily in the interval

$$\vartheta \in (-\pi/2, \pi/2) \text{ for } q(t_0) > 0$$

or

$$\vartheta \in (-\pi, -\pi/2) \cup (\pi/2, \pi] \text{ for } q(t_0) < 0 .$$

This lack of uniqueness is a straightforward consequence of the classical sampling theorem in section 3.1.4. In fact, for $M = 2$ there exist Dirac pulses δ in points $\pm \Omega/2$ in the spectrum of the continuous-time signal $x_c(t)$ corresponding to the carrier signal $q(t_n)$.

In order to compute the average pseudopower P_q of the sinusoidal modulation sequence notice first that

$$q(t_n) = \frac{1}{\sqrt{2}} \left(Q e^{j2\pi M^{-1}n} + Q^* e^{-j2\pi M^{-1}n} \right) \tag{4.22}$$

where

$$Q = |Q| e^{j\vartheta} .$$

and the complex conjugate to Q is denoted by Q^*. From equation (4.22) we get the following formula for the instantaneous pseudopower

$$\begin{aligned} p_{sq}(t_n) &= q^2(t_n) \tag{4.23} \\ &= \frac{1}{2} \left(2QQ^* + Q^2 e^{j4\pi M^{-1}n} + Q^{*2} e^{-j4\pi M^{-1}n} \right) . \end{aligned}$$

If we now compute the sequence average pseudopower P_q from formula (4.17) using equations (4.16) and (4.23), then for $M \geq 3$ the last two terms in equation (4.23) will cancel. In order to obtain a similar cancellation for $M = 1$ and $M = 2$, we can make use of the freedom in selecting parameter values $|Q|$ and ϑ. Parameter ϑ should be chosen in such a way that the following equality holds

$$Q^2 + Q^{*2} = 0 .$$

Thus, we get

$$\vartheta = \pm j\pi/4 .$$

Table 4.1 Elementary sinusoidal sequences

Sequence	Length M	Amplitude $\sqrt{2}|Q|$	Average pseudopower $P_q = M_1/M$
(1)	1	$\sqrt{2}$	1
$(1,-1)$	2	$\sqrt{2}$	1
$(1,-1,0)$	3	$2/\sqrt{3}$	2/3
$(1,0,-1,0)$	4	1	1/2
$(1,1,-1,-1)$	4	$\sqrt{2}$	1
$(1,1,0,-1,-1,0)$	6	$2/\sqrt{3}$	2/3

Under this assumption, we can generally write

$$P_q = \frac{1}{M} \sum_{n=0}^{M-1} q^2(t_n) = |Q|^2 , \qquad (4.24a)$$

i.e.

$$|Q| = \sqrt{P_q} . \qquad (4.24b)$$

We shall now consider all possible essentially different elementary sinusoidal sequences. Two sequences

$$\tilde{q} = (q_0, q_1, \ldots, q_{M-1}) \text{ and } \tilde{p} = (p_0, p_1, \ldots, p_{M-1})$$

are called **essentially equivalent**, denoted as

$$\tilde{q} \rightleftharpoons \tilde{p}$$

if the sequence $\tilde{q}$ can be obtained from the sequence $\tilde{p}$ by means of the following transformations of essential equivalence:

- circular shift

$$\tilde{p_{(m)}} = (p_m, p_{m+1}, \ldots, p_{M-1}, p_0, p_1, \ldots, p_{m-1}) ,$$

- mirror reflection

$$-\tilde{p} = (-p_0, -p_1, \ldots, -p_{M-1})$$

- or a combination of these operations.

Thus,
$$\tilde{q} \rightleftharpoons \tilde{p} \Leftrightarrow \bigvee_{0 \leq m \leq M-1} (p_{(m)}^{\sim} = \tilde{q}) \vee (-p_{(m)}^{\sim} = \tilde{q}) \ .$$

Sequences which are not essentially equivalent are called essentially different.

It can easily be checked that all possible essentially different elementary sinusoidal sequences covering one period of a sinusoidal signal are those listed in Table 4.1.

4.3 MULTIRATE SYSTEM CONCEPT

Notice that the process of sampling rate expansion (section 4.1.1) and the process of sampling rate compression (section 4.1.2) can be considered as particular cases of product modulation. In fact, instead of using the sampling rate expander in order to generate the signal $x_u(\tau_0 + nT)$ in Fig. 4.1 from the signal $x_o(\tau_\nu)$, we can apply a product modulator with an auxiliary input signal

$$\tilde{x}_o(\tau_0 + nT) = \begin{cases} x_o(\tau_\nu) & \text{for } n = \nu M \text{ and } \tau_\nu = \tau_0 + \nu T_o \\ \text{arbitrary} & \text{for } n \neq \nu M \ . \end{cases}$$
(4.25)

The carrier signal $q(t_n)$ is then composed of M-element modulation sequences of the form

$$\tilde{q} = (1, \underbrace{0, 0, \ldots, 0}_{M-1 \text{ zeros}}) \ .$$
(4.26)

For sampling rate compression, a similar corollary is obvious. It follows from a comparison of expression (4.11a) with expression (4.15a) in which $Q(m\Omega_o)$ is replaced by $1/Me^{-jm\Omega_o\tau_0}$. This value for $Q(m\Omega_o)$ follows from formula (3.19a) with t_0 replaced by τ_0 and $q(t_n)$ determined by (4.26).

Strictly speaking, product modulation with a carrier signal $q(t_n)$ composed of sequences (4.26) is equivalent first to a sampling rate compression and then to a sampling rate expansion, connected in cascade. This equivalence is illustrated in Fig. 4.6.

Heteromerous sampling, discussed in section 3.1.1, can be interpreted in a similar way. A modulation sequence consists in this case

74   Multirate systems

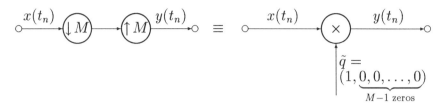

Fig. 4.6 Illustration of the fact that a cascade of a sampling rate compressor by a factor M with a sampling rate expander by a factor M is equivalent to a particular product modulator.

of values 1 and 0 only but elements equal to 1 represent sampling instants.

From the above discussion, we may conclude that it is reasonable to generalize the notion of multirate signal processing beyond the traditional sense of sampling rate alteration only. Introducing a generalized multirate system concept, based on a product modulation, sampling rate alteration and heteromerous sampling can be considered as particular cases of this general process.

Consider the system shown in Fig. 4.7. It is composed of two filters and a product modulator. The first filter with the transfer function $H_1(e^{j\omega T})$ is the so-called **premodulation filter**. It produces the effective signal $x_e(nT)$ from the input signal $x(t_{1n})$, $t_{1n} = t_{10} + nT$. The product modulator forms a signal $x_q(nT)$ by modulating a carrier signal $q(nT)$ by the effective signal $x_e(nT)$. On the basis of expression (4.15a), a spectrum of the signal $x_q(nT)$ is composed of translations of the effective signal $x_e(nT)$ spectrum. Thus, the second filter, the so-called **postmodulation filter** with the transfer function $H_2(e^{j\omega T})$, is necessary to reduce the signal $x_q(nT)$ spectrum to the desired bands of the output signal $y(t_{2n})$, $t_{2n} = t_{20} + nT$.

The system in Fig. 4.7 is called the **generalized multirate system** if

- the carrier signal $q(nT)$ is periodic with period T_o defined by equation (3.18) and the corresponding modulation sequence (4.14) contains at least one element equal to 0,

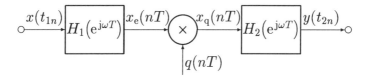

Fig. 4.7 Generalized multirate system scheme.

- continuous-time signals $x_c(t)$ and $y_c(t)$ represented by signals $x_e(nT)$ and $y(t_{2n})$ are integer L-band signals with respect to frequency Ω_o.

The former requirement follows from the fact that the system should realize heteromerous sampling or sampling rate alteration, whereas the latter condition means that the transfer functions $H_1(e^{j\omega T})$ and $H_2(e^{j\omega T})$ are under ideal circumstances defined by expression (4.9), possibly with different parameters const, ω_{l1}, ω_{l2}, $l = 1, 2, \ldots, L$ for each transfer function.

4.4 ANALYSIS OF MULTIRATE SYSTEMS

4.4.1 Systems with periodically varying parameters

The generalized multirate system defined in the previous section is linear but periodically time-varying with period T_o. Such a system cannot be described by a single time-independent transfer function as such a description is restricted to linear and shift-invariant systems only. In our case a more general analysis is required which can be derived from an approach based on a single resultant time-varying transfer function, proposed originally by Zadeh (1950) and Belevitch (1968) for the analysis of continuous-time periodically time-varying systems.

Consider first the system in Fig. 4.7 in a steady-state assuming, as in section 3.2.4, that the excitation signal $x(t_{1n})$ is of the form

$$x(t_{1n}) = X e^{p t_{1n}} \qquad (4.27a)$$

where X is a complex constant. The output signal $y(t_{2n})$ can have

76   Multirate systems

a more complicated waveform but it can always be expressed as

$$y(t_{2n}) = Y(z, t_{2n})e^{pt_{2n}} \tag{4.27b}$$

where $Y(z, t_{2n})$ is a complex-valued function; variables z and p are related by expression (3.39).

In a steady-state we can define the resultant time-varying transfer function

$$H = H(z, t_{2n}) = Y(z, t_{2n})/X . \tag{4.28a}$$

This transfer function is valid also in a general case, i.e. the following formula

$$Y(z, t_{2n}) = H(z, t_{2n})X(z) \tag{4.28b}$$

expresses the relation between the spectrum type time-varying function $Y(e^{j\omega T}, t_{2n})$ of the output signal $y(t_{2n})$ and the Fourier transform $X(e^{j\omega T})$ of the input signal $x(t_{1n})$.

The corresponding expression in time domain has the same form as formula (3.15c). Taking equation (4.28b) into account, we obtain

$$\begin{aligned} y(t_{2n}) &= \frac{1}{\Omega} \int_{-\Omega/2}^{\Omega/2} Y\left(e^{j\omega T}, t_{2n}\right) e^{j\omega t_{2n}} d\omega \\ &= \frac{1}{\Omega} \int_{-\Omega/2}^{\Omega/2} H\left(e^{j\omega T}, t_{2n}\right) X\left(e^{j\omega T}\right) e^{j\omega t_{2n}} d\omega . \end{aligned} \tag{4.29}$$

The output signal $y(t_{2n})$ can also be computed using the so-called **impulse response** $h = h(t_{2n}, t_{1m})$ representing the output signal with samples occurring in instants t_{2n}, corresponding to the unity impulse excitation in instant t_{1m}, i.e. to $x(t_{1n}) = \delta(t_{1n} - t_{1m})$ where

$$\delta(t_n) = \begin{cases} 1 & \text{for } t_n = 0 \\ 0 & \text{for } t_n \neq 0 . \end{cases}$$

Taking into account that the system is periodically time-varying with period T_o, we can write

$$h(t_{2n}, t_{1m}) = h(t_{2n} + \nu T_o, t_{1m} + \nu T_o) \tag{4.30}$$

where ν is an arbitrary integer. Exploiting the system linearity, the output signal can be computed as

$$y(t_{2n}) = \sum_{m=-\infty}^{\infty} h(t_{2n}, t_{1m})x(t_{1m}) . \tag{4.31}$$

Now comparing the right-hand sides of formulae (4.29) and (4.31) and using expression (3.15c), we finally get

$$H(z, t_{2n}) = \sum_{m=-\infty}^{\infty} h(t_{2n}, t_{1m}) e^{-p(t_{2n}-t_{1m})} \qquad (4.32a)$$

and

$$h(t_{2n}, t_{1m}) = \frac{1}{\Omega} \int_{-\Omega/2}^{\Omega/2} H\left(e^{j\omega T}, t_{2n}\right) e^{j\omega(t_{2n}-t_{1m})} d\omega \ . \qquad (4.32b)$$

Thus, on the basis of equality (4.30) and expression (4.32a), we conclude that the transfer function $H(e^{j\omega T}, t_{2n})$ is periodic in time with period T_o (determined by equation (3.18)). Hence, adapting expressions (3.19a) and (3.19b) we can write

$$H(z, t_{2n}) = \sum_{m=0}^{M-1} H_m(z) e^{jm\Omega_o t_{2n}} , \qquad (4.33a)$$

where

$$H_m(z) = \frac{1}{M} \sum_{n=0}^{M-1} H(z, t_{2n}) e^{-jm\Omega_o t_{2n}} , \qquad (4.33b)$$
$$m = 0, 1, \ldots, M-1 \ .$$

From expressions (4.33a) and (4.33b) it follows that the system considered can be uniquely described by M functions $H_m(z)$ called **conversion functions**. Function $H_m(z)$ is referred to as the mth order conversion function. The 0th order function $H_0(z)$ is also called the **main conversion function**.

Substituting decomposition (4.33a) into expression (4.29) we obtain

$$\begin{aligned} y(t_{2n}) &= \frac{1}{\Omega} \int_{-\Omega/2}^{\Omega/2} \sum_{m=0}^{M-1} H_m\left(e^{j\omega T}\right) e^{jm\Omega_o t_{2n}} X\left(e^{j\omega T}\right) e^{j\omega t_{2n}} d\omega \\ &= \sum_{m=0}^{M-1} \frac{1}{\Omega} \int_{-\Omega/2}^{\Omega/2} H_m\left(e^{j\omega T}\right) X\left(e^{j\omega T}\right) e^{j(\omega+m\Omega_o)t_{2n}} d\omega .(4.34) \end{aligned}$$

Using the following notation

$$Y_m\left(e^{j(\omega+m\Omega_o)T}\right) = H_m\left(e^{j\omega T}\right) X\left(e^{j\omega T}\right) \qquad (4.35)$$

and changing the integration variable from ω to $\omega - m\Omega_o$ results in the following decomposition of the output signal

$$y(t_{2n}) = \sum_{m=0}^{M-1} y_m(t_{2n}) \qquad (4.36)$$

with

$$\begin{aligned} y_m(t_{2n}) &= \frac{1}{\Omega} \int_{-\Omega/2-m\Omega_o}^{\Omega/2-m\Omega_o} Y_m\left(e^{j\omega T}\right) e^{j\omega t_{2n}} d\omega \\ &= \frac{1}{\Omega} \int_{-\Omega/2}^{\Omega/2} Y_m\left(e^{j\omega T}\right) e^{j\omega t_{2n}} d\omega \ . \end{aligned} \qquad (4.37)$$

Now changing again the order between summation and integration in expression (4.34), we get

$$y(t_{2n}) = \frac{1}{\Omega} \int_{-\Omega/2}^{\Omega/2} \sum_{m=0}^{M-1} Y_m\left(e^{j\omega T}\right) e^{j\omega t_{2n}} d\omega \ . \qquad (4.38)$$

Thus, the Fourier transform of the output signal $y(t_{2n})$ is equal to

$$Y\left(e^{j\omega T}\right) = \sum_{m=0}^{M-1} Y_m\left(e^{j\omega T}\right) \ . \qquad (4.39)$$

Notice that this is a different function from that defined by equation (4.28b). Indeed, comparing expressions (4.29) and (4.34) and using the notation of (4.35) we get

$$Y\left(e^{j\omega T}, t_{2n}\right) = \sum_{m=0}^{M-1} Y_m\left(e^{j(\omega+m\Omega_o)T}\right) e^{jm\Omega_o t_{2n}} \ , \qquad (4.40)$$

i.e. a different formula from (4.39).

Frequency ω in expression (4.35) corresponds to the input signal $x(nT)$, while frequencies $\omega + m\Omega_o$, $m = 0, 1, \ldots, M-1$, correspond to the output signal $y(t_{2n})$ or more precisely to its respective components $y_m(t_{2n})$ introduced in equation (4.36). By fitting arguments of conversion functions $H_m(e^{j\omega T})$, $m = 0, 1, \ldots, M-1$, to the frequencies of the respective output signal components $y_m(t_{2n})$, i.e. by moving these functions along the ω axis by $m\Omega_o$, respectively, the so-called **fitted conversion functions**

$$T_m\left(e^{pT}\right) = H_m\left(e^{(p-jm\Omega_o)T}\right) \ , \qquad (4.41)$$

$$m = 0, 1, \ldots, M-1 \ ,$$

Analysis of multirate systems 79

are obtained. Substituting equation (4.41) into equation (4.35) we get

$$Y_m\left(e^{j\omega T}\right) = T_m\left(e^{j\omega T}\right) X\left(e^{j(\omega - m\Omega_o)T}\right), \quad (4.42)$$
$$m = 0, 1, \ldots, M-1.$$

4.4.2 Construction of frequency characteristics

We shall now analyse the multirate system in Fig. 4.7. To this end the theory of time-varying discrete systems, presented in previous section can be used. The transmission properties of the considered multirate system can be described by the output signal Fourier transform $Y(e^{j\omega T})$. Taking into account equation (4.15a), the following relation can immediately be written

$$Y\left(e^{j\omega T}\right) =$$
$$H_2\left(e^{j\omega T}\right) \sum_{m=0}^{M-1} Q(m\Omega_o) H_1\left(e^{j(\omega - m\Omega_o)T}\right) X\left(e^{j(\omega - m\Omega_o)T}\right). \quad (4.43)$$

Thus we conclude that

$$Y_m\left(e^{j\omega T}\right) =$$
$$H_2\left(e^{j\omega T}\right) Q(m\Omega_o) H_1\left(e^{j(\omega - m\Omega_o)T}\right) X\left(e^{j(\omega - m\Omega_o)T}\right) \quad (4.44a)$$

and

$$T_m\left(e^{j\omega T}\right) = H_2\left(e^{j\omega T}\right) Q(m\Omega_o) H_1\left(e^{j(\omega - m\Omega_o)T}\right) \quad (4.44b)$$

while

$$H_m\left(e^{j\omega T}\right) = H_2\left(e^{j(\omega + m\Omega_o)T}\right) Q(m\Omega_o) H_1\left(e^{j\omega T}\right). \quad (4.44c)$$

Hence

$$H\left(e^{j\omega T}, t_{2n}\right) =$$
$$\left(\sum_{m=0}^{M-1} H_2\left(e^{j(\omega + m\Omega_o)T}\right) Q(m\Omega_o) e^{j\Omega_o t_{2n}}\right) H_1\left(e^{j\omega T}\right) \quad (4.45a)$$

and

$$Y\left(e^{j\omega T}, t_{2n}\right) =$$
$$\left(\sum_{m=0}^{M-1} H_2\left(e^{j(\omega + m\Omega_o)T}\right) Q(m\Omega_o) e^{j\Omega_o t_{2n}}\right) H_1\left(e^{j\omega T}\right) X\left(e^{j\omega T}\right). \quad (4.45b)$$

80 Multirate systems

In a particular case, i.e. if the postmodulation filter $H_2(e^{j\omega T})$ does not exist (e.g. for decimation), expressions (4.45a) and (4.45b) can be simplified

$$H\left(e^{j\omega T}, nT\right) = q(nT)H_1\left(e^{j\omega T}\right) , \qquad (4.46a)$$

$$Y\left(e^{j\omega T}, nT\right) = q(nT)H_1\left(e^{j\omega T}\right) X\left(e^{j\omega T}\right) . \qquad (4.46b)$$

In another important particular case, the input signal $x(nT)$ is just the effective signal $x_e(nT)$. This situation occurs, for example, in interpolation. In this case the premodulation filter can be omitted. Our present assumption means that the effective signal spectrum $X_e(e^{j\omega T})$ is restricted to the passbands of the premodulation filter transfer function, determined by formulae (4.6)–(4.9). Thus, in this case we can define the following modified conversion functions

$$\check{H}\left(e^{j\omega T}\right) = \begin{cases} H\left(e^{j\omega T}\right) & \text{for } \omega \in \bigcup_{k=-\infty}^{\infty} \bigcup_{l=1}^{L}[\omega_{l1} + k\Omega, \omega_{l2} + k\Omega] \\ & \cup [-\omega_{l2} + k\Omega, -\omega_{l1} + k\Omega] \\ 0 & \text{otherwise} \end{cases}$$

(4.47)

restricted to the respective passbands. These functions are referred to as the **cut conversion functions**. Consequently, we can introduce also the **fitted cut conversion functions**

$$\check{T}_m\left(e^{pT}\right) = \check{H}_m\left(e^{(p-jm\Omega_o)T}\right) . \qquad (4.48)$$

In the considered case, equation (4.42) can be rewritten with the fitted cut conversion functions. Thus

$$Y_m\left(e^{j\omega T}\right) = \check{T}_m\left(e^{j\omega T}\right) X_e\left(e^{j(\omega-m\Omega_o)T}\right) , \qquad (4.49)$$

$$m = 0, 1, \ldots, M-1 .$$

An important property of the set of fitted cut conversion functions is that for an arbitrary frequency ω at most one of them is not zero. Thus, the following resultant conversion function can be defined

$$T\left(e^{j\omega T}\right) = \sum_{m=0}^{M-1} \check{T}_m\left(e^{j\omega T}\right) . \qquad (4.50)$$

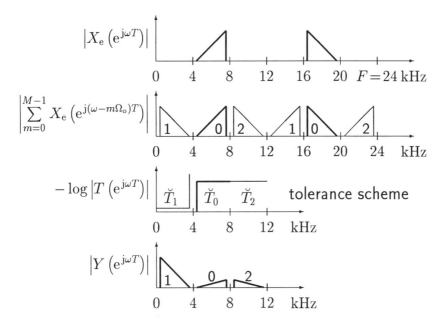

Fig. 4.8 Illustration of construction of frequency characteristics for a multirate system (multirate period $M = 3$).

The output signal spectrum can then be computed as

$$Y\left(e^{j\omega T}\right) = T\left(e^{j\omega T}\right) \sum_{m=0}^{M-1} X_e\left(e^{j(\omega - m\Omega_o)T}\right) . \qquad (4.51)$$

Using the resultant conversion function, the system attenuation, phase and group delay frequency characteristics can be determined as follows:

$$\alpha = 20\log_{10}\left|T^{-1}\left(e^{j\omega T}\right)\right| \text{ (dB)} \qquad (4.52\text{a})$$
$$\beta = \arg T\left(e^{j\omega T}\right) \text{ (rad)} \qquad (4.52\text{b})$$
$$\tau = -\frac{d\beta}{d\omega} \text{ (s)} , \qquad (4.52\text{c})$$

respectively.

To illustrate the above construction of frequency characteristics, consider a low-pass multirate system with sampling rate $F = 24$ kHz and multirate period $M = 3$. Assume that the effective signal

$x_e(nT)$ corresponds to an integer-band signal with respect to frequency $\Omega_o = 2\pi F/M$ with its spectrum restricted to the band $\omega \in [4,8]$ kHz. The construction of the frequency characteristic of this system is illustrated by means of the tolerance scheme in Fig. 4.8. This tolerance scheme is divided into three parts corresponding to fitted cut conversion functions $\check{T}_0$, $\check{T}_1$ and $\check{T}_2$ according to expressions (4.50) and (4.52a).

To end this section, it should be stressed, as a supplementary remark, that both filters, the premodulation filter and the postmodulation filter, can by themselves be periodically time-varying with a period T_o. In such a case, we assume that all their fitted conversion functions approximate ideal characteristics determined by equation (4.9), but the parameter const can be different for all the conversion functions and parameters ω_{l1}, ω_{l2}, $l = 1, 2, \ldots, L$, can also be different for both filters.

4.4.3 Equivalences for building-block connections

Building-block interconnections commonly occur in multirate systems. A number of simple equivalences for typically interconnected multirate building-blocks are summarized in Fig. 4.9 for sampling rate increase and in Fig. 4.10 for sampling rate decrease.

In Fig. 4.6, we illustrated the equivalence between a particular product modulator and a system composed of a sampling rate compressor by a factor M and a sampling rate expander by the same factor M, connected together in cascade. Now we shall investigate further equivalences for more general cascade connections of multirate building-blocks.

Two different ways of cascading a sampling rate expander by a factor L with a sampling rate compressor by a factor M are shown in Fig. 4.11. It can easily be shown from the definition of the sampling rate expander and the sampling rate compressor that the systems in Fig. 4.11 are equivalent if and only if factors L and M are relatively prime, i.e. if there exists no integer greater than 1 which is a common factor of both numbers L and M. The proof is illustrated in the bottom part of Fig. 4.11. If we first compress the

Analysis of multirate systems 83

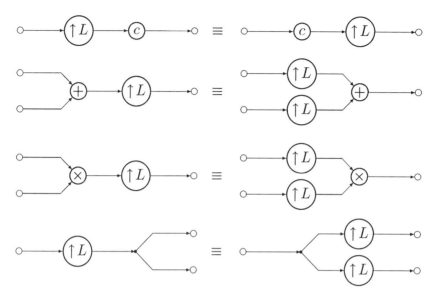

Fig. 4.9 Simple equivalences for interconnected sampling rate expanders.

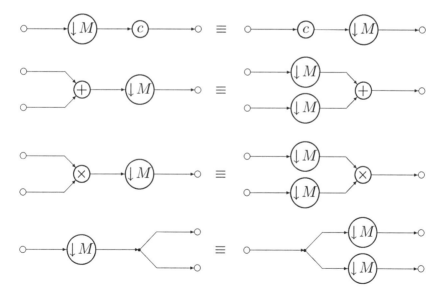

Fig. 4.10 Simple equivalences for interconnected sampling rate compressors.

84   *Multirate systems*

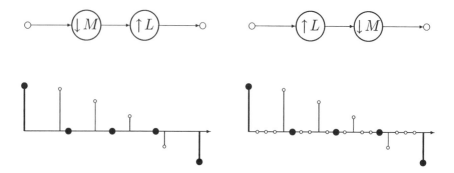

Fig. 4.11 Cascading a sampling rate expander with a sampling rate compressor.

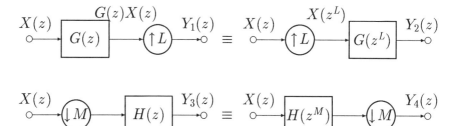

Fig. 4.12 Noble identities.

sampling rate by a factor M and then expand it by a factor L, we will get the sample stream shown on the left-hand side of Fig. 4.11. We will get the same stream by first expanding the sampling rate by a factor L and then compressing it by a factor M if and only if the next nonzero (nonexpanded) sample after a certain nonzero (nonexpanded) sample, obtained during the compression, is only the MLth sample. In this case M and L are indeed relatively prime.

Very useful also are the so-called **noble identities** illustrated in Fig. 4.12. To prove the first identity note from the left-hand side of the top of Fig. 4.12 that on the basis of equation (4.4b) we get

$$Y_1(z) = G(z^L)X(z^L)$$

which agrees with $Y_2(z)$ on the right-hand side of the top of Fig. 4.12, concluding the proof of the first identity.

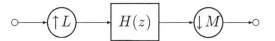

Fig. 4.13 Sampling rate alteration by a rational factor L/M.

The second identity can be proved using expression (4.12a). Note that

$$Y_4(z) = \frac{1}{M} \sum_{m=0}^{M-1} W_M^{m\tau_0/T} H\left((z^{1/M}W_M^m)^M\right) X(z^{1/M}W_M^m)$$

$$= \frac{1}{M} \sum_{m=0}^{M-1} W_M^{m\tau_0/T} H(z) X(z^{1/M}W_M^m)$$

$$= H(z) \frac{1}{M} \sum_{m=0}^{M-1} W_M^{m\tau_0/T} X(z^{1/M}W_M^m)$$

which is thus the same as $Y_3(z)$. This observation completes the proof.

Connecting in cascade an interpolator by a factor $\uparrow L$ (Fig. 4.1) and a decimator by a factor $\downarrow M$ (Fig. 4.3), we can achieve the sampling rate alteration by a rational factor L/M. The resulting arrangement is shown in Fig. 4.13. The transfer function $H(e^{j\omega T})$ represents a cascade of an interpolator filter in Fig. 4.1 and a decimator filter in Fig. 4.3 and in practice usually describes an appropriate low-pass filter.

4.4.4 Polyphase representation

For a stationary discrete-time system, equation (4.31) determining the output signal $y(t_{2n})$, $t_{2n} = t_{20} + nT$, reduces to the form

$$y(t_{2n}) = \sum_{m=-\infty}^{\infty} h(t_{20} + (n-m)T) x(t_{1m}) \qquad (4.53)$$

where $h(t_{2n})$ is the impulse response with samples occurring in instants t_{2n}, corresponding to the unity impulse excitation in instant $t_{1m} = t_{10} + mT$ and $x(t_{1n})$ is the input signal. Hence, for the transfer function

$$H\left(e^{j\omega T}\right) = Y\left(e^{j\omega T}\right) / X\left(e^{j\omega T}\right)$$

Multirate systems

we obtain from expression (4.32a)

$$\begin{aligned} H\left(e^{j\omega T}\right) &= e^{-j\omega(t_{20}-t_{10})}\mathcal{H}\left(e^{j\omega T}\right) \\ &= z^{-(t_{20}-t_{10})/T}\mathcal{H}(z) \end{aligned} \quad (4.54)$$

where

$$\begin{aligned} \mathcal{H}\left(e^{j\omega T}\right) &= \sum_{n=-\infty}^{\infty} h(t_{2n})e^{-j\omega nT} \\ &= \sum_{n=-\infty}^{\infty} h(t_{2n})z^{-n} \end{aligned} \quad (4.55)$$

is the conventional $\mathcal{Z}$ transfer function and $z = e^{j\omega T}$. In this particular (stationary) case, function $H(z)$ is simply a delayed by $t_{20} - t_{10}$ version of function $\mathcal{H}(z)$. This effect is of secondary importance and we will henceforth assume $t_{20} - t_{10} = 0$ for stationary systems.

Using equation (3.53b) we find the following polyphase decomposition

$$H(z) = \sum_{\nu=0}^{M-1} z^{-\nu} \mathcal{S}_\nu(z^M) \quad (4.56)$$

where

$$\mathcal{S}_\nu(z) = \sum_{n=-\infty}^{\infty} h(t_{2(Mn+\nu)})z^{-n} \ . \quad (4.57)$$

For nonrecursive, i.e. finite impulse response (FIR) transfer functions, this decomposition can be readily derived by inspection. However, using this approach for recursive, i.e. infinite impulse response (IIR) transfer functions, we obtain infinite polyphase sequences which must then be rewritten in closed form. This is inconvenient, and the approach described below may be used in this case.

Let

$$H(z) = \frac{\sum_{k=0}^{N_a} a_k z^{-k}}{1 - \sum_{l=1}^{N_b} b_l z^{-l}} = \frac{A(z)}{B(z)} \quad (4.58)$$

where $B(z)$ is a polynomial of degree

$$N_B = \max(N_a, N_b) \ .$$

Notice that $B(z)$ is a monic polynomial, i.e. the coefficient of the highest power term z^{N_B} is equal to one. The degree N_A of polynomial $A(z)$ cannot be greater than N_B, i.e. $N_A \leq N_B$. With some as yet unknown polynomial $C(z)$ of degree N_C, such that

$$B(z)C(z) = B_C(z^M) \qquad (4.59)$$

where $B_C(z^M)$ is a polynomial in z^M of degree

$$N_B + N_C = MN$$

with respect to variable z, function $H(z)$ may be rewritten as

$$\begin{aligned} H(z) &= \frac{A(z)C(z)}{B(z)C(z)} = \frac{E(z)}{B_C(z^M)} \\ &= \frac{\sum_{\nu=0}^{M-1} z^{-\nu} \mathcal{E}_\nu(z^M)}{z^{-MN} B_C(z^M)} \end{aligned} \qquad (4.60)$$

Comparing expressions (4.56) and (4.60), we conclude that in the IIR case

$$\mathcal{S}_\nu(z) = \frac{z^N \mathcal{E}_\nu(z)}{B_C(z)} \ . \qquad (4.61)$$

To find the unknown polynomial $C(z)$ notice that

$$B_C(z^M) = \text{const} \prod_{k=1}^{b} (z^M - z_k^M) \qquad (4.62)$$

where b and z_k, $k = 1, 2, \ldots, b$, are as yet unknown constants. Computing zeros z_{kl} of the polynomial $z^M - z_k^M$ in z we get

$$z_{kl} = z_k e^{j(2\pi l)/M} \ , \quad l = 0, 1, \ldots, M-1 \ . \qquad (4.63)$$

Thus, polynomial $B_C(z^M)$ can be expressed as

$$B_C(z^M) = \text{const} \prod_{k=1}^{b} \prod_{l=0}^{M-1} \left(z - z_k e^{j(2\pi l)/M} \right) \ . \qquad (4.64)$$

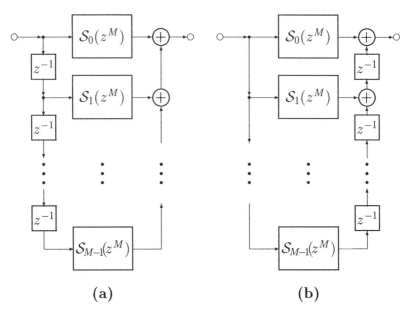

Fig. 4.14 Polyphase realization of transfer function $H(z)$ defined by equation (4.56): (a) type 1 structure, (b) type 2 structure.

Specifying now a real constant const in equation (4.64), we can accept

$$B_C(z^M) = \prod_{k=1}^{b} \prod_{l=0}^{M-1} \left(z e^{-j(2\pi l)/M} - z_k \right) . \tag{4.65}$$

Notice from equation (4.59) that

$$C(z) = \frac{B_C(z^M)}{B(z)} \tag{4.66}$$

is an unknown polynomial. Hence, all zeros of polynomial $B(z)$ must also be zeros of polynomial $B_C(z^M)$. Since constants b and z_k, $k = 1, 2, \ldots, b$, are as yet unspecified, we may now make our choice for them assuming that $b = N_B$ and

$$B(z) = \prod_{k=1}^{N_B} (z - z_k) . \tag{4.67}$$

Substituting expressions (4.65) and (4.67) into equation (4.66), we finally get

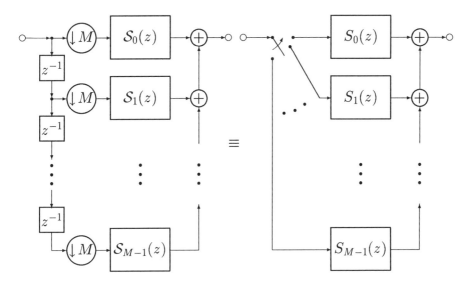

Fig. 4.15 Operationally efficient polyphase decimator structure.

$$C(z) = \prod_{l=1}^{M-1} \prod_{k=1}^{N_B} \left(ze^{-j(2\pi l)/M} - z_k\right)$$
$$= \prod_{l=1}^{M-1} B\left(ze^{-j(2\pi l)/M}\right) . \qquad (4.68)$$

A realization of transfer function $H(z)$, based on the polyphase decomposition (4.56) is shown in Fig. 4.14a. This is the so-called type 1 structure. Type 2 structure, illustrated in Fig. 4.14b, is obtained by changing the directions of all branches[*] in Fig. 4.14a.

The polyphase structures in Fig. 4.14 can be used for the realization of an operationally efficient decimator and interpolator. Basic realizations of these components, shown in Figs 4.1 and 4.3, are not efficient because filter $H(e^{j\omega T})$ operates with a higher sampling rate $F = \Omega/(2\pi)$. We can, however, realize this filter with polyphase components operating with a lower sampling rate $F_o = \Omega_o/(2\pi) = F/M$.

---

[*]Using Mason's rule, we conclude that changing directions of all branches does not change the transfer function.

90   *Multirate systems*

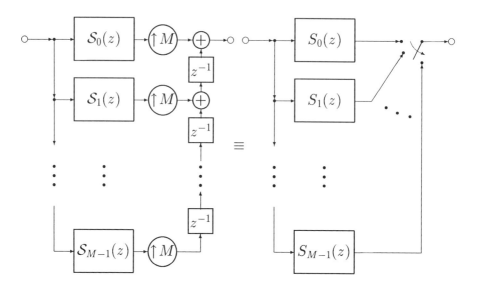

Fig. 4.16 Operationally efficient polyphase interpolator structure.

Efficient decimator structure can be obtained by realizing the transfer function $H(e^{j\omega T})$ in Fig. 4.3 by a type 1 polyphase structure modified using the second noble identity in Fig. 4.12, as shown on the left-hand side of Fig. 4.15. The tapped delay line with down-samplers $\downarrow M$ can be further replaced by an M-phase commutator, as illustrated on the right-hand side of Fig. 4.15. Conventional polyphase transfer functions $S_\nu(z)$, $\nu = 0, 1, \ldots, M-1$, must, however, be replaced by modified polyphase functions expressed according to equation (3.44) as

$$S_\nu(z) = z^{-\nu/M} \mathcal{S}_\nu(z) \ , \quad \nu = 0, 1, \ldots, M-1 \ . \tag{4.69}$$

Similarly we can obtain efficient interpolator structure but now we have to choose a type 2 polyphase structure and modify it using the first noble identity in Fig. 4.12. The resulting realization is shown on the left-hand side of Fig. 4.16. Again using modified polyphase transfer functions $S_\nu(z)$ given by equation (4.69), up-samplers $\uparrow M$ together with the delay line can be replaced by an M-phase commutator, as illustrated on the right-hand side of Fig. 4.16.

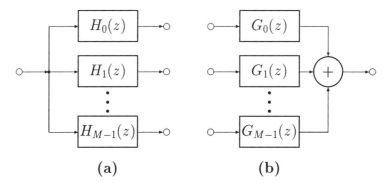

Fig. 4.17 Filter banks: (a) analysis filter bank, (b) synthesis filter bank.

4.5 FILTER BANKS

Filter banks are multi-input and/or multi-output frequency selective systems. The wave discrete-time system in Fig. 3.7 can serve as an example of a two-input and two-output filter bank, i.e. a bank of four filters.

The most typical filter banks of say M filters possess one input and M outputs, or conversely, they contain M inputs and one output (Fig. 4.17). The former are known as **analysis filter banks**, while the latter are referred to as **synthesis filter banks**. The frequency characteristics of individual filters in a filter bank are usually spread out along the frequency axis in such a way that the respective passbands cover the whole frequency range of interest. Typical frequency range partitions in filter banks are shown in Fig. 4.18. One of them corresponds to the so-called **uniform filter bank**, i.e. the bank composed of filters with uniformly distributed bandwidths, while the other illustrates a nonuniform (octave) filter bank. The latter is particularly useful in the subband coding of acoustic signals (speech and music) (Fettweis et al., 1990).

Subband coding (SBC) is a powerful technique for data compression in signals (Crochiere, Webber and Flanagan, 1976; Crochiere, 1981). It consists of splitting the signal into many subbands in such a way that:

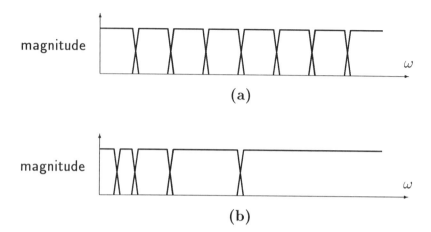

Fig. 4.18 Typical frequency range partitions in filter banks: (a) uniform filter bank, (b) nonuniform (octave) filter bank.

- the total number of samples remains unchanged

- the original signal can be reconstructed from the subband signals

- moreover, only a few subband signals are important in the reconstruction process, i.e. only these must be represented exactly, while the others are of secondary importance, i.e. these subband signals can be represented quite inexactly or they can even be omitted.

The SBC scheme is realized by an analysis filter bank connected in cascade with a synthesis filter bank. The analysis filter bank decomposes the signal spectrum in a number of directly adjacent frequency subbands. Then, the synthesis filter bank recombines these spectra and reconstructs the original signal. If the subbands are uniform, each of them will cover $1/M$th of the original signal spectrum only (Fig. 4.18a), and thus the signal decomposition can be connected with the sampling rate reduction by a factor M. Hence, the overall number of signal samples remains unchanged. In the process of signal reconstruction, the sampling rate must be increased again by a factor M. The resulting SBC system is shown in Fig. 4.19.

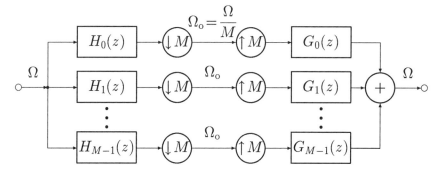

Fig. 4.19 Subband coding (SBC): signal decomposition into M subbands followed by signal reconstruction.

Notice that it consists of M multirate schemes of Fig. 4.7 connected in parallel. The cascades of the sampling rate compressor $\downarrow M$ and expander $\uparrow M$ that are present in the system of Fig. 4.19 are then replaced by the equivalent product modulators, as explained in Fig. 4.6.

Another common application of filter banks is in a **transmultiplexer**, i.e. the system converting time-division-multiplexed (TDM) signals[*] into frequency-division-multiplexed (FDM) signals and vice versa (Bellanger and Daguet, 1974; Bellanger, 1982; Fettweis, 1982; Vetterli, 1986b; Dąbrowski, 1987a, b). The synthesis filter bank is used for converting TDM signals into FDM signals and the analysis filter bank is used for the opposite operation.

4.5.1 Branching filter banks

A **branching filter bank**, or in other words a **directional filter bank**, is a pair of complementary filters forming a two-channel analysis filter bank $(H, \bar{H})$ or a two-channel synthesis filter bank $(G, \bar{G})$ as shown in Fig. 4.20. Typically one filter, e.g. H or G, is a low-pass filter with a passband width of approximately $\Omega/4$, while the other filter, i.e. $\bar{H}$ or $\bar{G}$, has a complementary high-pass frequency response (Fig. 4.21). Thus the frequency band partition

---

[*]This signal multiplexing technique is known in communications as the pulse-coded modulation (PCM).

94 Multirate systems

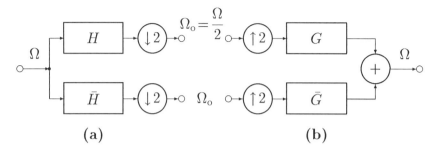

Fig. 4.20 Branching filter banks: (a) analysis filter bank $(H, \bar{H})$, (b) synthesis filter bank $(G, \bar{G})$.

is symmetrical. In such a case, the sampling rate can be reduced by a factor of 2 when splitting the signal into two subbands, i.e. in the analysis filter bank, and it can be increased by a factor of 2 when composing the signal from subbands in the synthesis filter bank (Fig. 4.20).

Using branching filter banks, the simplest (two-channel) SBC scheme can be realized, as illustrated in Fig. 4.22. The input signal $x(t_{1n})$ is first passed through the analysis filter bank and partitioned into two subband signals $x_{e0}(nT)$ and $x_{e1}(nT)$. We assume that the respective filter transfer functions $H = H_0(z)$ and $\bar{H} = H_1(z)$ correspond to low-pass and to high-pass frequency responses, respectively. Then the subband signals are down-sampled by a factor of 2. Assuming that the filters in the analysis bank are realized using switched-capacitor circuits, analog-to-digital converters and encoders are then inserted in each subband channel. These components are omitted from Fig. 4.22 for simplicity. Each subband signal is individually encoded, exploiting its importance for signal reconstruction (i.e. energy level, perceptual importance, etc.) (Jayant and Noll, 1984). The receiver decoders (also omitted from Fig. 4.22) then produce approximations of the original subband signals which are up-sampled by a factor of 2 again. These signals, denoted in Fig. 4.22 as $x_{q0}(nT)$ and $x_{q1}(nT)$, are then filtered with filters $G = G_0(z)$ and $\bar{G} = G_1(z)$ whose outputs are added, yielding the final output signal $y(t_{2n})$. Signal $y(t_{2n})$ represents the reconstruction of the original input signal $x(t_{1n})$.

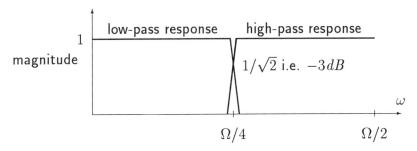

Fig. 4.21 Typical frequency responses of individual filters in a branching filter bank.

As illustrated in Fig. 4.6, the cascade of a sampling rate compressor by a factor of 2 with sampling rate expander by a factor of 2, used in the SBC scheme in Fig. 4.22, is equivalent to the product modulator with modulation sequence $\tilde{q} = (1,0)$. Taking this equivalence into account, the following input–output relation for the analysed SBC scheme can be directly obtained from expression (4.43)

$$Y\left(e^{j\omega T}\right) =$$

$$\sum_{k=0}^{1} G_k\left(e^{j\omega T}\right) \sum_{m=0}^{1} Q(m\Omega_o) H_k\left(e^{j(\omega - m\Omega_o)T}\right) X\left(e^{j(\omega - m\Omega_o)T}\right) . \quad (4.70)$$

From equation (3.19a) we conclude that in our case

$$Q(m\Omega_o) = \frac{1}{2} \quad \text{for} \quad m = 0 \text{ and } 1 .$$

Also taking into account that $\Omega_o T = \pi$, we obtain the following expression for the modified $\mathcal{Z}$ transforms

$$\begin{aligned} Y(z) &= \frac{1}{2} \sum_{k=0}^{1} G_k(z)\left[H_k(z)X(z) + H_k(-z)X(-z)\right] \\ &= \frac{1}{2}\left[G_0(z)H_0(z) + G_1(z)H_1(z)\right]X(z) \\ &+ \frac{1}{2}\left[G_0(z)H_0(-z) + G_1(z)H_1(-z)\right]X(-z) . \quad (4.71) \end{aligned}$$

This equation can be compactly expressed as

96 Multirate systems

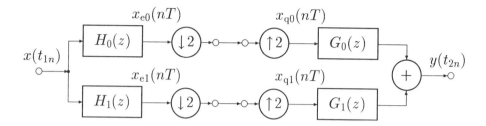

Fig. 4.22 Two-channel subband coding scheme.

$$Y(z) = D(z)X(z) + A(z)X(-z) \qquad (4.72a)$$

where

$$D(z) = \frac{1}{2}[G_0(z)H_0(z) + G_1(z)H_1(z)] \qquad (4.72b)$$

and

$$A(z) = \frac{1}{2}[G_0(z)H_0(-z) + G_1(z)H_1(-z)] \ . \qquad (4.72c)$$

The first term in equation (4.72a) describes the distortion caused by the signal transfer in the SBC scheme, while the second term is due to an aliasing effect caused by product modulation (equivalent to sampling rate alteration down and up by a factor of 2). Therefore, functions $D(z)$ and $A(z)$ are called the **distortion transfer function** and the **aliasing conversion function**, respectively.

A matrix representation of equation (4.72a) is given by

$$\begin{aligned} Y(z) &= \begin{bmatrix} D(z) & A(z) \end{bmatrix} \begin{bmatrix} X(z) \\ X(-z) \end{bmatrix} \\ &= \frac{1}{2} \begin{bmatrix} G_0(z) & G_1(z) \end{bmatrix} \begin{bmatrix} H_0(z) & H_0(-z) \\ H_1(z) & H_1(-z) \end{bmatrix} \begin{bmatrix} X(z) \\ X(-z) \end{bmatrix}. \end{aligned} \qquad (4.73)$$

As the product modulator (equivalent to a down-sampler connected in cascade with an up-sampler) is a linear but time-varying component, the analysed SBC scheme is a linear but, in general, time-varying system. However, it is possible to choose the analysis and synthesis filter banks in such a way that the aliasing is cancelled and the resulting SBC system is linear and time-invariant. To this

end, we need to ensure that
$$A(z) \equiv 0$$
for all input signals. There exist various solutions of above condition but one of the simplest is given by
$$G_0(z) = cH_1(-z) \tag{4.74a}$$
and
$$G_1(z) = -cH_0(-z) \tag{4.74b}$$
where c is an arbitrary (real) constant.

If above relations hold, then expression (4.72a) reduces to
$$Y(z) = D(z)X(z) \tag{4.75a}$$
with
$$D(z) = \frac{c}{2}\left[H_0(z)H_1(-z) - H_0(-z)H_1(z)\right] . \tag{4.75b}$$
Thus, the corresponding relation for Fourier transforms is given by
$$Y\left(e^{j\omega T}\right) = D\left(e^{j\omega T}\right) X\left(e^{j\omega T}\right) . \tag{4.75c}$$
If $D(z)$ is an all-pass function, i.e.
$$\left|D\left(e^{j\omega T}\right)\right| = \text{const} \neq 0$$
for all frequencies ω, then the SBC scheme is without magnitude distortion*. In such a case, the SBC scheme is said to be **magnitude preserving**.

If transfer function $D(z)$ has linear phase, i.e.
$$\beta(\omega) = \arg D\left(e^{j\omega T}\right) = -\tau\omega + \beta_0$$
then
$$\arg Y\left(e^{j\omega T}\right) = \arg X\left(e^{j\omega T}\right) - \tau\omega + \beta_0$$
and the group delay
$$\tau = -\frac{d\beta(\omega)}{d\omega}$$

---
*We certainly neglect a constant gain which can possibly be different from 1.

is constant. In the particular case for $\beta_0 = 0$, not only the group delay but also the 'absolute' delay is constant (equal to τ) and the SBC scheme is without phase distortion. Thus, in this case it is said to be **phase preserving**.

If an aliasing-free SBC scheme has no magnitude and phase distortion, then it is said to be a **perfect reconstruction** (PR) scheme (Vetterli, 1986a). In such a case

$$D(z) = \text{const} \cdot z^{-\tau/T} \qquad (4.76)$$

resulting in

$$Y(z) = \text{const} \cdot z^{-\tau/T} X(z)$$

or in time domain

$$y(t_{2n}) = \text{const} \cdot x(t_{2n} - \tau) = \text{const} \cdot x(t_{1n})$$

with $t_{20} = t_{10} + \tau$, i.e. $t_{2n} = t_{1n} + \tau$.

4.5.2 Quadrature-mirror filter banks

In this section we shall show that an attempt to eliminate magnitude and phase distortion in a two-channel subband coding scheme discussed in previous section leads to **power complementary filter banks** or in other words to **quadrature-mirror filter** (QMF) **banks** whose low-pass and high-pass responses fulfil the following power complementary condition*

$$\left|H_0\left(e^{j\omega T}\right)\right|^2 + \left|H_1\left(e^{j\omega T}\right)\right|^2 = 1 \qquad (4.77\text{a})$$

or more generally

$$\left|H_0\left(e^{j\omega T}\right)\right|^2 + \left|H_1\left(e^{j\omega T}\right)\right|^2 = \text{const} > 0 \ . \qquad (4.77\text{b})$$

Analytical extension of property (4.77a) from the unit circle to the whole z plane is given by†

$$H_0(z)H_0(z^{-1}) + H_1(z)H_1(z^{-1}) = 1 \ . \qquad (4.77\text{c})$$

---

*Traditionally, systems fulfilling condition (4.77a) or (4.77b) only approximately are also called QMF banks (Johnston, 1980).

†We certainly assume that $H_0(e^{j\omega T})$ and $H_1(e^{j\omega T})$ are rational functions with real coefficients.

Comparing expressions (4.77a)–(4.77c) with expression (3.81) we conclude that QMF banks are examples of pseudolossless (paraunitary) systems.

Starting the design of the SBC scheme with a suitable low-pass filter $H_0(z)$ of the analysis filter bank, the simplest way for realization of the second (high-pass) filter $H_1(z)$ is the choice

$$H_1(z) = H_0(-z) \ . \tag{4.78}$$

This is in fact the earliest known QMF bank (Croisier, Esteban and Galand, 1976; Esteban and Galand, 1977). Substituting relation (4.78) into equations (4.74a) and (4.74b), we obtain in this case

$$G_0(z) = cH_0(z) \tag{4.79a}$$

and

$$G_1(z) = -cH_1(z) \ . \tag{4.79b}$$

On the unit circle, equation (4.78) can be written as

$$H_1\left(e^{j\omega T}\right) = H_0\left(e^{j(\omega - \Omega/2)T}\right) \ .$$

Thus

$$\left|H_1\left(e^{j\omega T}\right)\right| = \left|H_0\left(e^{j(\Omega/2-\omega)T}\right)\right|$$

for rational functions with real coefficients. Substituting $\omega = \Omega/4$ into the above equation, we conclude that

$$\left|H_0\left(e^{j(\Omega/4)T}\right)\right| = \left|H_1\left(e^{j(\Omega/4)T}\right)\right| = \frac{1}{\sqrt{2}} \ .$$

Thus, the magnitude responses of both filters are symmetrical with respect to $\omega = \Omega/4$ and cross each other for this frequency at the relative magnitude value of $1/\sqrt{2}$, i.e. at the attenuation of 3 dB (cf. Fig. 4.21).

Substituting equation (4.78) into expression (4.77c) gives

$$H_0(z)H_0(z^{-1}) + H_0(-z)H_0(-z^{-1}) = 1 \ . \tag{4.80a}$$

Thus using the notation

$$T(z) = H_0(z)H_0(z^{-1}) \tag{4.80b}$$

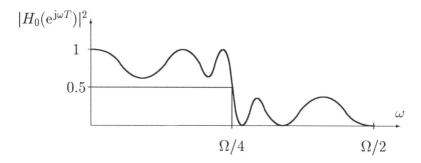

Fig. 4.23 Magnitude-squared characteristic of a low-pass QMF (pseudolossless) filter.

we get
$$T(z) + T(-z) = 1 \ . \qquad (4.80c)$$

Filters with transfer function $T(z)$ fulfilling condition (4.80c) are known as **half-band filters** (Vaidyanathan, 1993). Hence, a filter with the transfer function $H_0(z)H_0(z^{-1})$ is a half-band filter. Moreover, it has a non-negative amplitude response. Therefore it is called the **valid half-band filter** (Fliege, 1993).

On the unit circle, equation (4.80a) reduces to
$$\left|H_0\left(e^{j(\omega+\Omega/4)T}\right)\right|^2 = 1 - \left|H_0\left(e^{j(\Omega/4-\omega)T}\right)\right|^2 ,$$
which shows that the magnitude-squared characteristic exhibits symmetry with respect to the middle point of this curve at $\omega = \Omega/4$ and $|H_0(e^{j\omega T})|^2 = 0.5$, i.e. the passband and the stopband waveforms are symmetrical to each other (Fig. 4.23).

With the choice of condition (4.78) and the aliasing cancellation constraints (4.74a), (4.74b) chosen with constant $c = 2$, the distortion transfer function is given by
$$D(z) = H_0^2(z) - H_1^2(z) = H_0^2(z) - H_0^2(-z) \ . \qquad (4.81)$$

Let us now restrict the following considerations to a QMF bank with **finite impulse response** (FIR) **filters**, which, as we know, can be designed with exactly linear phase. If $H_0(z)$ is a linear phase function, then the distortion transfer function $D(z)$ given by (4.81) also has a linear phase, thereby eliminating phase distortion. In order to analyse the magnitude distortion, notice that a low-pass linear phase FIR transfer function can be expressed by

$$H_0(z) = z^{-N/2}R(z) \tag{4.82}$$

where $R(e^{j\omega T})$ is a zero-phase function, i.e. a real function for all frequencies ω, and N is the filter order or the minimum number of delays ($N+1$ being the number of filter coefficients, i.e. the number of samples of the impulse response) (Rabiner and Gold, 1975). Substituting equation (4.82) into expression (4.81) and using the fact that

$$-1 = e^{j\pi} = e^{j(\Omega/2)T} ,$$

gives

$$D\left(e^{j\omega T}\right) =$$
$$e^{-jN\omega T}\left(R^2\left(e^{j\omega T}\right) - (-1)^N R^2\left(e^{j(\omega-\Omega/2)T}\right)\right)$$

$$= e^{-jN\omega T}\left(\left|H_0\left(e^{j\omega T}\right)\right|^2 - (-1)^N \left|H_0\left(e^{j(\omega-\Omega/2)T}\right)\right|^2\right) . \tag{4.83}$$

To achieve acceptable magnitude distortion, order N cannot be an even number because in such a case expression (4.83) reduces to zero at point $\omega = \Omega/4$, resulting in severe magnitude distortion. Thus, we have to choose N as an odd number, so that

$$\begin{aligned}D\left(e^{j\omega T}\right) &= e^{-jN\omega T}\left(\left|H_0\left(e^{j\omega T}\right)\right|^2 + \left|H_0\left(e^{j(\omega-\Omega/2)T}\right)\right|^2\right) \\ &= e^{-jN\omega T}\left(\left|H_0\left(e^{j\omega T}\right)\right|^2 + \left|H_1\left(e^{j\omega T}\right)\right|^2\right) .\end{aligned} \tag{4.84}$$

Note from equation (4.84) that if condition (4.77a) (or more generally (4.77b)) were fulfilled, $D(e^{j\omega T})$ would reduce to a pure delay $\tau = NT$ and, on the basis of equation (4.76), the analysed SBC scheme would realize a perfect reconstruction (PR) system. Unfortunately, as we shall show below, even if N is odd, condition (4.77a) or (4.77b) can only be approximately fulfilled with FIR filters related by expression (4.78), except from a quite special and unpractical case determined by relations (4.91a) and (4.91b). However, the residual magnitude distortion can be kept arbitrarily small if we choose an appropriately high filter order (Johnston, 1980; Jain and Crochiere, 1984). Indeed, if the filter order is sufficiently high, then $H_0(e^{j\omega T})$ can be a very good low-pass filter with $|H_0(e^{j\omega T})| \approx 1$ in the passband, $|H_0(e^{j\omega T})| \approx 0$ in the stopband and a very narrow transition band. Then $H_1(e^{j\omega T})$ will be a very good high-pass filter with its passband coinciding with the stopband of $H_0(e^{j\omega T})$ and vice versa. As a result, $|D(e^{j\omega T})| \approx 1$ for almost all frequencies. Significant distortion can only occur in the transition band of both filters, which can, however, be made very narrow and optimized in shape to achieve approximately constant magnitude $|D(e^{j\omega T})|$.

We shall now consider a quite special structure for the analysed SBC scheme – one that is very efficient, especially with **infinite impulse response** (IIR) filters. This structure is obtained by realizing the analysis and synthesis filters in a polyphase form.

Using condition (4.78) and expression (4.56) with $M = 2$, we can represent functions $H_0(z)$ and $H_1(z)$ with their polyphase components as follows:

$$H_0(z) = \frac{1}{2}\left(S_0(z^2) + z^{-1}S_1(z^2)\right) \qquad (4.85\text{a})$$

and

$$H_1(z) = \frac{1}{2}\left(S_0(z^2) - z^{-1}S_1(z^2)\right) \qquad (4.85\text{b})$$

where polyphase transfer functions are denoted by

$$\frac{1}{2}S_0(z) \text{ and } \frac{1}{2}S_1(z)$$

for simplicity of further expressions. Assuming that conditions (4.74a) and (4.74b) are fulfilled with constant $c = 2$, we get

$$G_0(z) = S_0(z^2) + z^{-1}S_1(z^2) \qquad (4.86\text{a})$$

and

$$G_1(z) = -S_0(z^2) + z^{-1}S_1(z^2) \; . \qquad (4.86\text{b})$$

Equations (4.85a), (4.85b) and (4.86a), (4.86b) can be rewritten in matrix form as

$$\begin{bmatrix} H_0(z) \\ H_1(z) \end{bmatrix} = \frac{1}{2}\begin{bmatrix} 1 & 1 \\ 1 & -1 \end{bmatrix}\begin{bmatrix} S_0(z^2) \\ z^{-1}S_1(z^2) \end{bmatrix} \qquad (4.87\text{a})$$

and

$$\begin{bmatrix} G_0(z) & G_1(z) \end{bmatrix} = \begin{bmatrix} z^{-1}S_1(z^2) & S_0(z^2) \end{bmatrix}\begin{bmatrix} 1 & 1 \\ 1 & -1 \end{bmatrix} , \qquad (4.87\text{b})$$

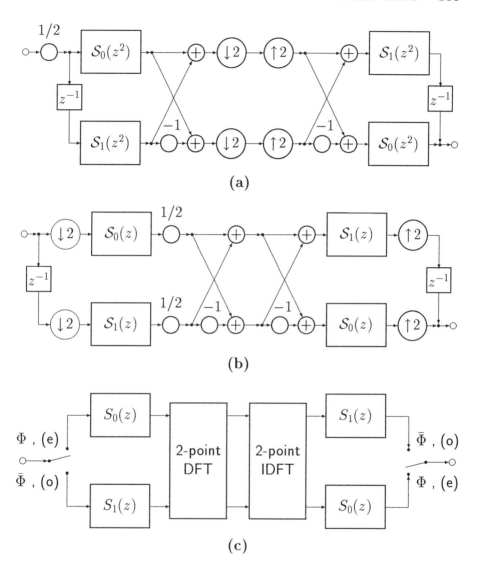

Fig. 4.24 Two-channel SBC polyphase scheme: (a) original scheme, (b) and (c) modified efficient schemes.

respectively.

The corresponding SBC scheme is illustrated in Fig. 4.24a. It can be modified using noble identities (Fig. 4.12) and following a similar procedure to that used in section 4.4.4 to derive efficient realizations for the decimator and interpolator, as indicated in Fig. 4.24b. The obtained structure is more efficient because the polyphase components S_0 and S_1 now operate with half the previous sampling rate. Moreover, similarly to the transformations shown in Figs 4.15 and 4.16, the delays, down-samplers and up-samplers in Fig. 4.24b can be replaced by two switches which are controlled with a two-phase clock denoted in Fig. 4.24c as Φ, $\bar{\Phi}$ or (e), (o) (e for an even phase and o for an odd phase, respectively). Conventional polyphase transfer functions $\mathcal{S}_0(z)$ and $\mathcal{S}_1(z)$ must, however, be replaced by modified polyphase functions given, according to equation (4.69), by

$$S_0(z) = \mathcal{S}_0(z)$$

and

$$S_1(z) = z^{-1/2}\mathcal{S}_1(z) \ .$$

Notice also that the first matrix on the right-hand side of equation (4.87a) (together with the coefficient 1/2) and the second matrix on the right-hand side of equation (4.87b) realize a two-point DFT and a two-point IDFT, respectively. Taking all these observations into account results in the structure illustrated in Fig. 4.24c.

Substituting functions (4.85a)–(4.86b) into equation (4.72b), we conclude that the distortion transfer function in this case (i.e. if condition (4.78) holds) is given by

$$D(z) = S_0(z^2)S_1(z^2) = z^{-1}\mathcal{S}_0(z^2)\mathcal{S}_1(z^2) \ . \qquad (4.88)$$

From equation (4.88) we note that the magnitude distortion can be perfectly eliminated in this case[*] if

$$\left|\mathcal{S}_0\left(e^{j\omega T}\right)\right| = \left|\mathcal{S}_1\left(e^{j\omega T}\right)\right|^{-1} \ . \qquad (4.89)$$

In a particular case, if, for example, $H_0(z)$ is an FIR function, so also are functions $H_1(z)$, $\mathcal{S}_0(z)$, $\mathcal{S}_1(z)$ and $D(z)$. Thus, condition (4.89) can be fulfilled in this case if and only if both polyphase transfer functions $\mathcal{S}_0(z)$ and $\mathcal{S}_1(z)$ are all-pass functions, i.e.

$$\left|\mathcal{S}_0\left(e^{j\omega T}\right)\right| = \left|\mathcal{S}_1\left(e^{j\omega T}\right)\right| = 1 \ . \qquad (4.90)$$

---

[*]With the additional simplifying assumption that the overall SBC scheme gain is equal to 1.

However, an FIR function is an all-pass function only if it represents a pure delay. Thus we can write

$$S_0(z) = z^{-n_0}$$

and

$$S_1(z) = z^{-n_1}$$

for some integers n_0 and n_1 corresponding to delays $\tau_0 = n_0 T$ and $\tau_1 = n_1 T$. Consequently, transfer functions of the analysis branching filter are in this case restricted to

$$H_0(z) = z^{-2n_0} + z^{-2n_1-1} \qquad (4.91a)$$

and

$$H_1(z) = z^{-2n_0} - z^{-2n_1-1} \ . \qquad (4.91b)$$

Such filters are very primitive and usually insufficiently selective. Therefore, using practical FIR filters fulfilling condition (4.78), we can only minimize magnitude distortion by numerical optimization, as already mentioned above.

Now we consider a much more general situation in which at least one of the polyphase component filters in the analysed SBC scheme in Fig. 4.24 is an IIR filter. In this case, condition (4.90) is not necessary but still strongly recommended. First of all, if condition (4.90) is fulfilled, the SBC is internally pseudolossless, since all its building-blocks, i.e. DFT, IDFT and all-pass polyphase sections $S_0(z)$, $S_1(z)$ are pseudolossless. Hence, such an SBC scheme possesses all the marvellous robustness properties related to internal pseudolosslessness such as low passband sensitivity, great dynamic range and stability (cf. section 3.2.4). Furthermore, there exists a powerful methodology for the design of good filters composed of all-pass sections (Gazsi, 1985; Vaidyanathan, 1993). Consequently, choosing functions $S_0(z)$ and $S_1(z)$ to be all-pass functions is one of the best methods to implement QMF filter banks free from aliasing and magnitude distortion.

At the end of this section, we will discuss an approach to the design of perfect reconstruction QMF filter banks. Notice that the power complementary conditions (4.77c) and (4.80a) can also be satisfied without fulfilling condition (4.78), i.e. even if expressions (4.79a) and (4.79b) do not hold, although we would certainly suppress the aliasing component in equation (4.72a). Thus we still assume that the anti-aliasing constraints (4.74a) and (4.74b) are satisfied. With the above assumptions in mind, the following causal solution for FIR filters of odd order N can be found (Smith and Barnwell, 1984)

$$H_1(z) = z^{-N} H_0(-z^{-1}) , \quad (4.92a)$$

$$G_0(z) = 2z^{-N} H_0(z^{-1}) = -2H_1(-z) \quad (4.92b)$$

and

$$G_1(z) = 2H_0(-z) . \quad (4.92c)$$

The above equations show that the transfer function $H_1(z)$ is derived from function $H_0(z)$ not only by a low-pass/high-pass transformation $z \to -z$ but also by causal time inversion. Transfer function $G_0(z)$ follows from function $H_0(z)$ by time inversion. On the other hand, function $G_1(z)$ follows from function $H_0(z)$ by low-pass/high-pass transformation.

Time inversion in FIR filters can be easily implemented in the switched-capacitor technique. In the IIR case, however, we should process signals blockwise using two memory buffer blocks. If samples stored in one block are processed forwards, then those in the second block are processed backwards and vice versa. This technique can be realized in digital networks but it is not suitable for SC realizations.

Substituting relations (4.92a)–(4.92c) and (4.80a) into equation (4.71) results in $A(z) \equiv 0$ and $D(z) = z^{-N}$. Thus, this is a perfect reconstruction scheme with the distortion transfer function representing a pure delay NT. In order to design the component filters in this scheme, we start from a low-pass filter $H_0(z)$ satisfying condition (4.80a). Design of this filter consists in finding a valid half-band filter $T(z)$ fulfilling requirement (4.80c) (this filter can be designed by, for example, using the McClellan-Parks algorithm) (McClellan and Parks, 1973), and then, in performing the spectral factorization (4.80b). The other filters in this SBC scheme are then derived from relations (4.92a)–(4.92c).

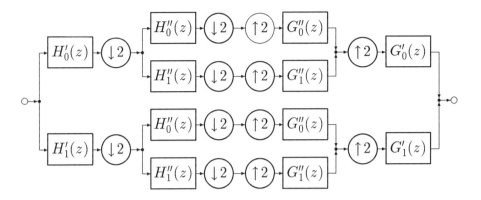

Fig. 4.25 Four-channel tree-structured uniform SBC scheme equivalent to the non-tree structure in Fig. 4.19 for $M = 4$.

Notice that in the FIR case, a generalized factorization given by

$$z^{-N}T(z) = H_0(z)H_1(-z) \tag{4.93}$$

can also be used, where N, as before, is an odd integer. We assume that $z^{-N}T(z)$ is a polynomial in z^{-1}. A part of its zeros can be arbitrarily assigned to the transfer function $H_0(z)$, the rest to the function $H_1(-z)$. If we then choose

$$G_0(z) = 2H_1(-z) \tag{4.94a}$$

and

$$G_1(z) = -2H_0(-z) \tag{4.94b}$$

for filters in the synthesis filter bank, we achieve a perfect reconstruction SBC scheme again. This is the so-called **biorthogonal SBC scheme** (Fliege, 1993, 1995).

In present case, function $T(z)$ can be an arbitrary half-band filter (i.e. it needs no longer to be valid). Moreover, filters $H_0(z)$ and $H_1(z)$ need no longer be power complementary. However, they can be linear phase filters. Indeed, if function $T(z)$ has linear phase, then it can always be factorized into two functions $H_0(z)$ and $H_1(-z)$ which also represent linear phase filters. This is advantageous in the coding of digitized versions of subband signals.

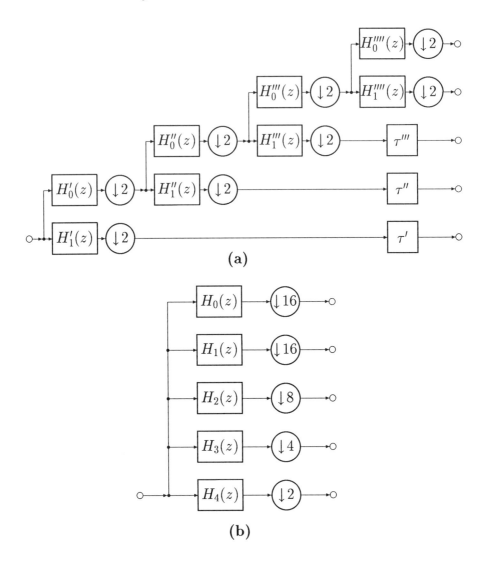

Fig. 4.26 Octave analysis filter bank: (a) four-stage tree-structured realization, (b) equivalent non-tree realization.

4.5.3 Multichannel filter banks

The two-channel SBC polyphase structure shown in Fig. 4.24 can be directly generalized to an arbitrary number of channels M (Bel-

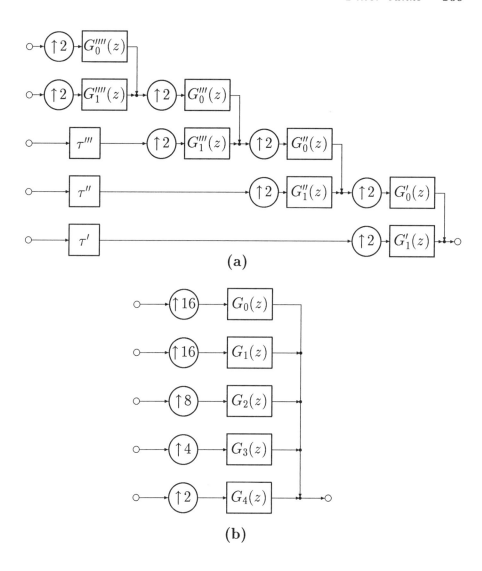

Fig. 4.27 Octave synthesis filter bank: (a) four-stage tree-structured realization, (b) equivalent non-tree realization.

langer, Bonnerot and Coudreuse, 1976; Vetterli, 1987; Vaidyanathan, 1993). The resulting polyphase analysis filter bank has the same structure as the polyphase decimator on the right-hand side of Fig. 4.15 but with the output summers replaced by an M-point

110 *Multirate systems*

DFT. Similarly, the respective polyphase synthesis filter bank structure is obtained from the polyphase interpolator on the right-hand side of Fig. 4.16 by replacing the signal branching point at the input to this system by an M-point IDFT. Unfortunately, such structures are not suitable for realization with switched-capacitor circuits unless $M = 2$, as for $M > 2$ they require complex arithmetic operations.

However, various interesting, efficient multi-channel filter bank structures, which are appropriate for implementation in switched-capacitor technique, can be obtained by multiple use of the exclusively two-channel QMF banks described in the previous section. The resulting systems have tree-type structures and are therefore called **tree-structured filter banks**.

For example, the maximally decimated uniform analysis filter bank on the left-hand side of Fig. 4.19 can be realized for $M = 2^N$ by an N-stage tree-structured filter bank composed by the two-band analysis QMF banks of Fig. 4.20a. The QMF banks in each stage split the signal of every subband into two new subbands and compress the sampling rate by a factor of 2. An inverse procedure of N-stage subband composing and two-fold signal up-sampling in each stage realizes the corresponding M-channel synthesis filter bank containing the two-band synthesis QMF banks of Fig. 4.20b. An illustrative SBC structure of this kind is shown for $M = 4$ in Fig. 4.25. If all of the component two-channel SBC schemes are perfect reconstructions then the complete M-channel scheme is also a perfect reconstruction.

A similar technique can be used for the realization of SBC schemes with octave filter banks. Illustrative four-stage structures for the analysis filter bank and the synthesis filter bank realizing the frequency range partition of Fig. 4.18b are shown in Figs 4.26a and 4.27a, respectively. In Figs 4.26b and 4.27b the corresponding equivalent non-tree structures are indicated. In order to achieve the perfect reconstruction property for the overall SBC system realized as the cascade of filter banks in Figs 4.26a and 4.27a, assuming that the component two-channel SBC schemes are perfect reconstructions, supplementary delays $2\tau'$, $2\tau''$ and $2\tau'''$ must be added in the

SBC structure in order to compensate for the differences of delays in particular branches. The system in Fig. 4.26 realizes the so-called **multiresolution decomposition** of the input signal. The output signal of the uppermost branch represents a 'coarse' (low-band) approximation of the original signal. Signals in other branches represent band-limited corrections with bandwidths growing in consecutive octave intervals. Summing these signals successively from the top to the bottom but starting with the uppermost signal, i.e. with the minimum resolution signal approximation, results in a series of signal approximations with growing resolution level. This type of signal representation is directly related to the **wavelet transformation** concept (Mallat, 1989; Chui, 1992; Vetterli and Herley, 1992).

4.6 MULTIRATE WAVE DISCRETE-TIME ARRANGEMENTS

The wave discrete-time systems introduced in section 3.2 represent a class of networks which are closely related to classical lossless filter circuits, typically inserted between resistive terminations. Thus, every wave discrete-time system can be derived from a specific classical filter network called its reference circuit (Fig. 1.1a).

The most interesting are certainly pseudolossless (or in other words paraunitary) discrete-time systems (cf. section 3.2.4), i.e. those whose reference circuits are lossless. Such discrete-time systems can also be derived directly, i.e. without an electrical network theoretical background and exploiting the correspondence with the reference circuit. The 'direct' approach has been developed and used particularly for the design of digital realizations (Deprettere and Dewilde, 1980; Vaidyanathan, 1993). However, the correspondence between the wave discrete-time arrangement and its reference circuit has turned out to be very fruitful. This is especially the case for switched-capacitor realizations, which are inherently analog electrical networks. Therefore, we follow in this book the original derivation of pseudolossless discrete-time systems based on the reference circuit concept.

It should also be stressed that there exists a class of switched-

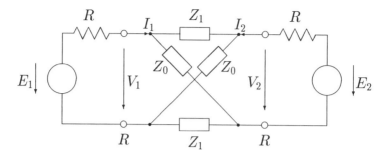

Fig. 4.28 Symmetrical lattice reference circuit.

capacitor circuits, the so-called VIS-SC circuits containing voltage inverter switches (Fettweis, 1979a, b) whose structures exactly reflect structures of their reference circuits. This is due to a one-to-one correspondence between components of both structures. The easiest and, at the same time, the most systematic way of designing VIS-SC circuits is first to design the reference circuit, and then to make use of this correspondence.

There exist, generally, two types of reference circuit structure: **lattice structure** and **ladder structure**. To both of them there are corresponding interesting multirate arrangements which are suitable for various switched-capacitor realizations (Brückmann, 1986; Dąbrowski, 1988a). Although ladder structure is much more important in classical continuous-time applications, symmetrical lattice structure (Fig. 4.28) seems to be advantageous in many aspects of discrete-time signal processing, including the multirate case. However, for multirate ladder wave discrete-time structures, a specific design methodology has also been developed (Fettweis and Nossek, 1982; Dąbrowski and Fettweis, 1987; Dąbrowski, 1988b) that in some applications leads to substantial advantages compared to the standard design for the same signal-processing purpose.

4.6.1 Multirate lattice arrangements

A symmetrical lattice two-port reference circuit inserted between two equal resistive terminations R is illustrated in Fig. 4.28. In typical filtering applications, lattice impedances Z_0 and Z_1 (known

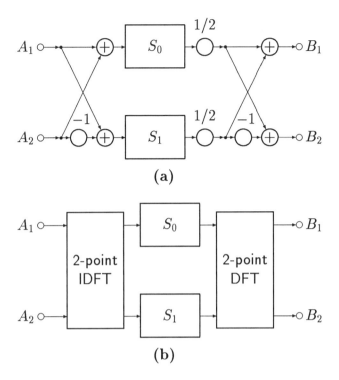

Fig. 4.29 Wave discrete-time lattice scheme: (a) detailed structure, (b) compact representation.

also as **canonic impedances**) realize reactances. In this case, the so-called **canonic reflectances** defined as

$$S_0 = \frac{Z_0 - R}{Z_0 + R} \quad \text{and} \quad S_1 = \frac{Z_1 - R}{Z_1 + R} \qquad (4.95)$$

are both all-pass functions.

The scattering transfer functions defined by equation (3.79) are then given by

$$S_{11} = S_{22} = \frac{S_0 + S_1}{2} \qquad (4.96\text{a})$$

and

$$S_{12} = S_{21} = \frac{S_0 - S_1}{2} \, . \qquad (4.96\text{b})$$

Using equations (4.96a) and (4.96b), expression (3.79) can be rewrit-

114 Multirate systems

ten as follows

$$B_1 = \frac{1}{2}S_0(A_1 + A_2) + \frac{1}{2}S_1(A_1 - A_2) \quad (4.97a)$$

and

$$B_2 = \frac{1}{2}S_0(A_1 + A_2) - \frac{1}{2}S_1(A_1 - A_2) \quad (4.97b)$$

or in matrix form

$$\begin{bmatrix} B_1 \\ B_2 \end{bmatrix} = \frac{1}{2}\begin{bmatrix} 1 & 1 \\ 1 & -1 \end{bmatrix}\begin{bmatrix} S_0 & \\ & S_1 \end{bmatrix}\begin{bmatrix} 1 & 1 \\ 1 & -1 \end{bmatrix}\begin{bmatrix} A_1 \\ A_2 \end{bmatrix}. \quad (4.98)$$

A realization corresponding to above equation is represented in Fig. 4.29. It can be simplified in particular cases as shown in Fig. 4.30. For example, for input signal $A_2 \equiv 0$, we obtain a branching analysis filter bank with $H = S_{11}$ and $\bar{H} = S_{21}$ (Fig. 4.30a). On the other hand, assuming that signal B_2 is not used, results in a branching synthesis filter bank $G = S_{11}$ and $\bar{G} = S_{12}$ (Fig. 4.30b). The simplest, i.e. single-input/single-output case, e.g. for signal A_1 chosen as input and signal $2B_1$ used as output, is indicated in Fig. 4.30c.

Although for functions $S_0 = S_0(\psi)$ and $S_1 = S_1(\psi)$, ψ being the reference circuit complex frequency defined by equation (3.54), arbitrary all-pass functions can be chosen, the most important is the so-called **bireciprocal** case (Fettweis, 1986) characterized by relations

$$S_0(1/\psi) = \pm S_0(\psi) \quad (4.99a)$$

and

$$S_1(1/\psi) = \mp S_1(\psi) \quad (4.99b)$$

where either both upper signs or both lower signs have to be chosen. We choose the first alternative because the other one gives essentially equivalent results. Furthermore, notice that due to relation (3.54), replacement of frequency ψ by $1/\psi$ amounts to the replacement of variable z by $-z$ in equations (4.99a) and (4.99b). Thus we conclude that function S_0 is even and function S_1 is odd in variable z. Hence, we may write

$$S_0(\psi) = \mathcal{S}_0(z^2) \quad (4.100a)$$

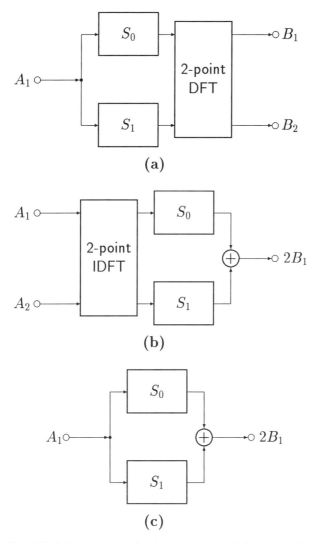

Fig. 4.30 Simplified discrete-time lattice structures: (a) analysis branching filter bank, (b) synthesis branching filter bank, (c) single-input/single-output filter.

and
$$\mathcal{S}_1(\psi) = z^{-1}\mathcal{S}_1(z^2) \ . \tag{4.100b}$$

Substituting equations (4.100a), (4.100b) into expressions (4.96a), (4.96b) and comparing the results with formulae (4.85a)–(4.86b),

116 *Multirate systems*

we conclude that in the bireciprocal case, both canonic reflectances play the role of two polyphase components for transfer functions

$$2S_{11} = 2S_{22} = 2H_0(z) = G_0(z)$$

and also for transfer functions

$$2S_{12} = 2S_{21} = 2H_1(z) = -G_1(z)$$

but in the latter case the transfer function $S_1(z^2)$ must be used with a negative sign. Taking also into account that Feldtkeller relations (3.81) are equivalent to power complementary conditions (4.77a)–(4.77c), we conclude that the bireciprocal lattice arrangements of Fig. 4.30a and Fig. 4.30b realize analysis and synthesis IIR QMF filter banks, respectively. Hence, they can be used in the aliasing and magnitude distortion free SBC schemes of Fig. 4.24. However, in the synthesis filter bank of Fig. 4.30b, reflectance S_0 must be replaced by reflectance S_1 and vice versa, according to the difference in sign between equations (4.96b) and (4.86b).

On the other hand, the single-input/single-output network version in Fig. 4.30c can be used in the bireciprocal case to implement the polyphase decimator and polyphase interpolator of Fig. 4.15 and Fig. 4.16, respectively, for $M = 2$.

Notice from the above considerations that the design of symmetrical lattice arrangements consists in obtaining two all-pass canonic reflectances S_0 and S_1 (or eventually two canonic reactances Z_0 and Z_1). There exists an efficient numerical procedure for solving this problem (Gazsi, 1985). For digital realizations, it is implemented in the Menthor Graphics Filter Architect software package, based on the original FALCON program (Gazsi, 1984).

Hereafter, we summarize the main steps of the procedure for computation of two all-pass canonic reflectances S_0 and S_1. We start with the solution of the approximation problem resulting in a desired characteristic function $K(\psi)$ that we assume to be an odd rational function given by

$$K(\psi) = h/f \ , \tag{4.101}$$

$h = h(\psi)$ and $f = f(\psi)$ being relatively prime polynomials. The characteristic function $K(\psi)$ is related to the filter normalized transfer function $H(\psi)$ by

$$K(\psi)K(-\psi) + 1 = \frac{1}{H(\psi)H(-\psi)} \qquad (4.102)$$

and to the filter attenuation α (defined for 'real' frequency $\psi = \mathrm{j}\phi$) by

$$\alpha = 10\lg\left(|K(\mathrm{j}\phi)|^2 + 1\right) \text{ (dB)} . \qquad (4.103)$$

Then we determine two Hurwitz polynomials $g' = g'(\psi)$ and $g'' = g''(\psi)$ such that

$$h(\psi) + f(\psi) = g'(-\psi)g''(\psi) . \qquad (4.104)$$

The above factorization is clearly always feasible. Finally, the canonical reflectances are computed by

$$S_0 = \frac{g'(-\psi)}{g'(\psi)} \quad \text{and} \quad S_1 = \frac{g''(-\psi)}{g''(\psi)} \qquad (4.105\mathrm{a})$$

for an even polynomial h and odd polynomial f and by

$$S_0 = \frac{g'(-\psi)}{g'(\psi)} \quad \text{and} \quad S_1 = -\frac{g''(-\psi)}{g''(\psi)} \qquad (4.105\mathrm{b})$$

in the opposite case.

For SC realizations with voltage inverter switches (VIS circuits), we can overcome the problem of computation of canonic reflectances although we have to determine canonic reactances Z_0 and Z_1. This is due to a one-to-one correspondence between the reference circuit elements and components of the resulting VIS-SC circuit. However, the above procedure can certainly also be used. Canonic reactances Z_0 and Z_1 are then given by

$$Z_0 = R\frac{1 + S_0}{1 - S_0} \qquad (4.106\mathrm{a})$$

and

$$Z_1 = R\frac{1 + S_1}{1 - S_1} . \qquad (4.106\mathrm{b})$$

They can be realized using standard methods for reactance synthesis, e.g. the Richards, Foster and/or Kuroda transformation method (Richards, 1948; Ozaki and Ishii, 1958; Belevitch, 1968; Dąbrowski, 1995). The only restriction is the requirement for solely grounded elements (cf. section 8.2).

4.6.2 Multirate ladder arrangements

Pseudolossless discrete-time systems which are suitable for multirate SC applications can also be derived from ladder reference circuits. Notice that even if we implement a lattice, or the more general polyphase structure of Fig. 4.14, particular elements such as all-pass sections or general polyphase components following from the decomposition (4.60) can be realized using ladder-type structures. For example, using VIS-SC circuits, the canonic reflectances S_0 and S_1 will be realized if we simply synthesize the canonic reactances Z_0 and Z_1. Classical synthesis procedures such as the Cauer, Foster, Richards and Kuroda methods lead to ladder-type structures (Belevitch, 1968).

Under the conception of **multirate ladder arrangements**, however, we understand the ladder realization of a complete system (interpolator, decimator or generalized multirate system of Fig. 4.7).

The ladder reference circuit approach is in some sense more general than that based on the use of lattice reference circuits because ladder filters need not be symmetric. As we shall show in section 8.3, this fact can be exploited to improve filtering discrimination properties without an increase in filter order (Fettweis and Nossek, 1982; Dąbrowski and Fettweis, 1987; Dąbrowski, 1988b).

On the other hand, multirate ladder filters cannot be decomposed into polyphase components and therefore they must operate with the higher sampling rate. Hence, they are less efficient in SBC schemes compared to equivalent lattice structures. If we, however, consider a single interpolator or decimator, then the polyphase structure does not help much. Although individual polyphase branches operate with a lower sampling rate, the overall hardware complexity (due to the existence of parallel polyphase branches) is approximately the same as for ladder structure of the same order. The latter can, however, exhibit better filtering discrimination and can, in the end, be beneficial for certain applications.

5

Systems with nonuniformly sampled signals

Nonuniform sampling of signals has gained a wide acceptance in many aspects of signal processing. For example, a technique of seamless changing of sampling rate can be used for data acquisition when performing transient analysis (McConnell, 1995). We can start to sample quickly to capture the transient. Afterwards, a slow sampling can be used to cope with a steady-state.

Periodic nonuniform sampling, in this book referred to as **heteromerous sampling**, is a particularly interesting technique. It consists in the repetitive use of a sequence with nonuniformly spaced sampling times (Fig. 3.3).

Heteromerous sampling is used in, for example, radar systems to realize moving-target-indicator filters in order to remove stationary (DC) components from the Doppler signals contained in the return echoes of moving targets (Roy and Lowenschuss, 1970). A special heteromerous sampling technique can be used to stabilize dynamic systems (Dąbrowski, 1981, 1983). Recently, a particular heteromerous time discretization has been proposed for sampling and stable reconstruction (interpolation) of band-unlimited signals (Vaidyanathan, 1995).

In switched-capacitor filters, heteromerous sampling has another interesting application: an SC filter can be driven with a heteromerous clock in order to shift its frequency response (Hurst, 1991; Korzec and Tounsi, 1994). This method makes possible very slight shifts, compared with ordinary control of the sampling rate with uniform sampling. It can thus be used for precise tuning and/or error correction in the frequency response of monolithic SC filters,

120 Systems with nonuniformly sampled signals

caused by capacitor ratio inaccuracies.

5.1 NONUNIFORM SAMPLING ANALYSIS

5.1.1 Commensurate sampling

Heteromerous sampling is closely related to multirate signal processing, discussed in Chapter 4. In order to show this relation, consider the general heteromerous sampling process illustrated in Fig. 3.3. Let $T_1, T_2, \ldots, T_M > 0$ be the corresponding individual sampling periods. The average sampling period can then be computed as

$$T = \frac{1}{M} \sum_{m=1}^{M} T_m = T_o/M \qquad (5.1)$$

where T_o is the overall period of the heteromerous sampling, given by equation (3.41).

As shown in section 3.1.6, the resulting stream of samples can be split into polyphase components (Fig. 3.4) that represent M uniformly sampled signals with period T_o.

In a particular case, all individual sampling periods $T_1, T_2, \ldots, T_M$ can be multiples of some elementary period (e.g. the system clock period) T_c, i.e.

$$T_m = M_m T_c \quad \text{for} \quad m = 1, 2, \ldots, M \ ,$$

M_m being arbitrary positive integers. Thus, the overall period of the sampling sequence is, in this case, given by

$$T_o = \sum_{m=1}^{M} T_m = N T_c \qquad (5.2)$$

where

$$N = \sum_{m=1}^{M} M_m \ . \qquad (5.3)$$

Such a technique is referred to as **heteromerous commensurate sampling**. It can be interpreted as nonuniform down-sampling of some signal sampled originally with rate $F_c = 1/T_c$. If this down-sampling operation is followed by up-sampling consisting of placing zero-valued samples in instants corresponding to samples just withdrawn (i.e. by the preceding down-sampling), then this process is equivalent to the product modulation of Fig. 4.5 with the modulation sequence composed of elements equal to 0 or 1, placed in the appropriate order. Values 1 correspond to samples left after down-sampling and values 0 to those which had to be withdrawn. Thus, the product modulation theory developed in sections 4.2, 4.3 and 4.4 is valid also in this case. Consequently, the commensurate sampling can be described by the system in Fig. 4.7, when coupled with filtering.

For a commensurately sampled signal $x(t_n)$, the Dirichlet transform $\mathcal{D}\{x(t_n)\}$ defined by equation (3.43) is equivalent to the $\mathcal{Z}$ transform $X(z)$ where variable $z = e^{pT_c}$ is defined with respect to elementary period T_c.

5.1.2 Nonuniform polyphase representation

Analysis of general heteromerous sampling is much more complicated than that for commensurate sampling, discussed in the previous section. Considerations in the present case can, however, be simplified by the assumption (which, as a matter of fact, is valid in most practical cases) that heteromerous sampling is an external operation to the system actually processing the signal. This assumption simply means that the heteromerous stream of samples is first equalized by appropriate shifts of samples in order to obtain their uniform distribution in time with the average sampling rate $F = 1/T$, according to equation (5.1). Such a signal occurs at the input of the system that actually operates with uniformly spaced samples and can, therefore, be a time-invariant system. It can thus be described by a common transfer function that will be denoted by $H(z)$. Appropriate samples of the output signal can then be shifted again in order to get a heteromerously sampled output signal. An illustrative arrangement based on this idea is presented in Fig. 5.1.

The system of Fig. 5.1 can be realized using efficient polyphase decomposition, discussed in section 4.4.4. In a particular case of

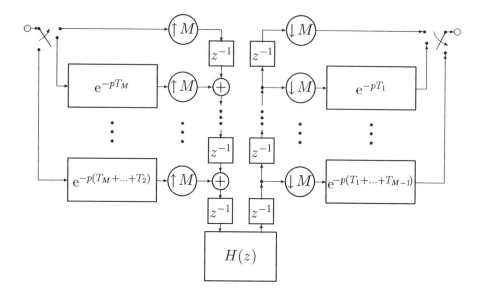

Fig. 5.1 System with heteromerously sampled signals at input and output.

nonuniform sampling rate alteration, with the additional assumption that sampling by the lower rate is uniform, structures appropriate for a polyphase nonuniform decimator and a polyphase nonuniform interpolator follow directly from those structures shown on the right-hand side of Figs 4.15 and 4.16, respectively. The only difference is a nonuniformly operating M-phase switch and nonuniform delays in the polyphase branches. Examples of such structures for a nonuniform decimator and a nonuniform interpolator are shown in Fig. 5.2a and b, respectively.

The idea of a polyphase representation can also be used for the analysis of the general heteromerously sampled system of Fig. 5.1. It is, in fact, enough to notice that the decimator of Fig. 5.2a, redrawn in Fig. 5.3a, realizes the first (0th) polyphase component $Y_0(e^{j\omega T_\circ})$ of the output signal $y(t_n)$. The next component $Y_1(e^{j\omega T_\circ})$ can be realized by the system in Fig. 5.3b and the last $(M-1)$th component $Y_{M-1}(e^{j\omega T_\circ})$ by the system in Fig. 5.3c.

The Dirichlet transform of the input signal $x(t_n)$ is given by expression (3.45). Similarly, we get also the Dirichlet transform of

Nonuniform sampling analysis 123

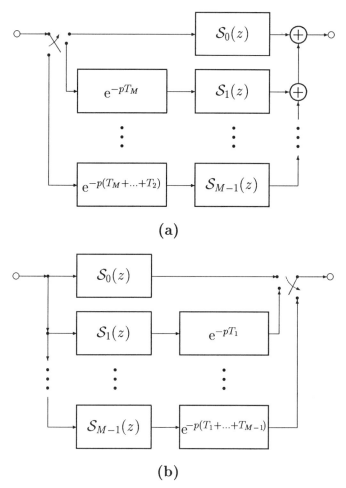

Fig. 5.2 Heteromerous sampling rate alteration arrangements: (a) decimator, (b) interpolator.

the output signal $y(t_n)$

$$\mathcal{D}\{y(t_n)\} = \sum_{\nu=0}^{M-1} Y_\nu(z) = \sum_{\nu=0}^{M-1} e^{-pt_\nu} \mathcal{Y}_\nu(z) \tag{5.4}$$

where instants t_ν, $\nu = 0, 1, \ldots, M - 1$, are defined by equation (3.42). Let us introduce the following notation for the polyphase transfer functions in the system of Fig. 5.3a

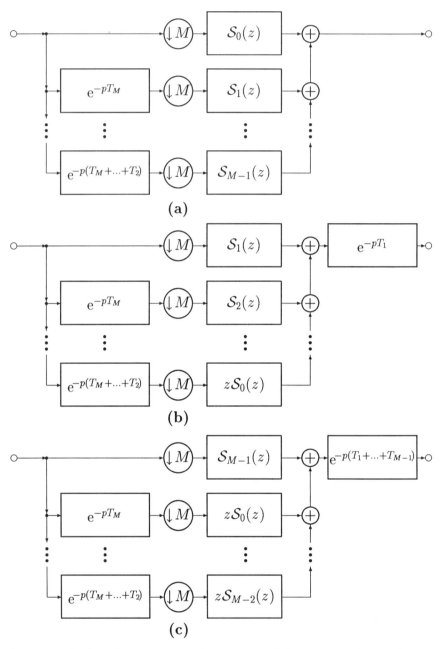

Fig. 5.3 Polyphase decomposition of a system with heteromerously sampled signals at input and output: (a), (b) and (c) realization of νth polyphase component of the output signal $y(t_n)$ for $\nu = 0$, 1 and $M - 1$, respectively.

$$S_0(z) = S_0(z) \tag{5.5a}$$
$$S_1(z) = e^{-pT_M}\mathcal{S}_1(z) \tag{5.5b}$$
$$S_2(z) = e^{-p(T_M+T_{M-1})}\mathcal{S}_2(z) \tag{5.5c}$$
$$\vdots \qquad \vdots$$
$$S_{M-1}(z) = e^{-p(T_M+T_{M-1}+\ldots+T_2)}\mathcal{S}_{M-1}(z) \tag{5.5d}$$

where variable z is defined with respect to period T_o. By inspection of Fig. 5.3a, we obtain the following expression for 0th polyphase component $Y_0(e^{j\omega T_o})$ of the output signal $y(t_n)$

$$\begin{aligned}
Y_0(z) &= S_0(z)X_0(z) + \sum_{\nu=1}^{M-1} S_\nu(z)X_{M-\nu}(z) \\
&= \mathcal{S}_0(z)e^{-pt_0}\mathcal{X}_0(z) + \sum_{\nu=1}^{M-1} \mathcal{S}_\nu(z)e^{-p(T_o+t_0)}\mathcal{X}_{M-\nu}(z) \\
&= e^{-pt_0}\left(\mathcal{S}_0(z)\mathcal{X}_0(z) + z^{-1}\sum_{\nu=1}^{M-1}\mathcal{S}_\nu(z)\mathcal{X}_{M-\nu}(z)\right) \\
&= e^{-pt_0}\mathcal{Y}_0(z) \ . \tag{5.6a}
\end{aligned}$$

Similarly we obtain also other polyphase components $Y_\nu(e^{j\omega T_o})$ for $\nu = 1, \ldots, M-1$

$$\begin{aligned}
Y_1(z) &= \Big(\mathcal{S}_1(z)e^{-pt_0}\mathcal{X}_0(z) + \sum_{\nu=2}^{M-1}\mathcal{S}_\nu(z)e^{-p(T_o+t_0)}\mathcal{X}_{M+1-\nu}(z) \\
&\quad + z\mathcal{S}_0(z)e^{-p(T_o+t_0)}\mathcal{X}_1(z)\Big)e^{-pT_1} \\
&= e^{-pt_1}\Big(\mathcal{S}_1(z)\mathcal{X}_0(z) + z^{-1}\sum_{\nu=2}^{M-1}\mathcal{S}_\nu(z)\mathcal{X}_{M+1-\nu}(z) \\
&\quad + \mathcal{S}_0(z)\mathcal{X}_1(z)\Big) \\
&= e^{-pt_1}\mathcal{Y}_1(z) \ . \tag{5.6b} \\
Y_2(z) &= \Big(\mathcal{S}_2(z)e^{-pt_0}\mathcal{X}_0(z) + \sum_{\nu=3}^{M-1}\mathcal{S}_\nu(z)e^{-p(T_o+t_0)}\mathcal{X}_{M+2-\nu}(z) \\
&\quad + z\sum_{\nu=0}^{1}\mathcal{S}_\nu(z)e^{-p(T_o+t_0)}\mathcal{X}_{1-\nu}(z)\Big)e^{-p(T_1+T_2)} \\
&= e^{-pt_2}\Big(\mathcal{S}_2(z)\mathcal{X}_0(z) + z^{-1}\sum_{\nu=3}^{M-1}\mathcal{S}_\nu(z)\mathcal{X}_{M+2-\nu}(z)
\end{aligned}$$

126 *Systems with nonuniformly sampled signals*

$$+ \sum_{\nu=0}^{1} \mathcal{S}_\nu(z)\mathcal{X}_{2-\nu}(z)\Bigg)$$
$$= e^{-pt_2}\mathcal{Y}_2(z) \ . \tag{5.6c}$$

$$\vdots \qquad \vdots$$

$$Y_{M-1}(z) = \Bigg(\mathcal{S}_{M-1}(z)e^{-pt_0}\mathcal{X}_0(z)$$
$$+ z\sum_{\nu=0}^{M-2} \mathcal{S}_\nu(z)e^{-p(T_\circ+t_0)}\mathcal{X}_{M-1-\nu}(z)\Bigg)e^{-p(T_1+T_2+\ldots+T_{M-1})}$$
$$= e^{-pt_{M-1}}\Bigg(\mathcal{S}_{M-1}(z)\mathcal{X}_0(z) + \sum_{\nu=0}^{M-2} \mathcal{S}_\nu(z)\mathcal{X}_{M-1-\nu}(z)\Bigg)$$
$$= e^{-pt_{M-1}}\mathcal{Y}_{M-1}(z) \ . \tag{5.6d}$$

Substituting the above polyphase components into equation (5.4) results in the Dirichlet transform of the output signal $y(t_n)$. Assuming in turn an input signal of the form

$$x(t_n) = e^{j\omega t_n}$$

i.e. substituting $p = j\omega$ and $z = e^{j\omega T_\circ}$, results in the steady-state response for a sinusoidal input.

5.2 SHIFTING THE FREQUENCY RESPONSE

An interesting technique for shifting the frequency response of switched-capacitor filters by nonuniform sampling was proposed by Hurst (1991) and then developed by Korzec and Tounsi (1994).

Consider first a frequency-selective SC filter (a resonator, or reversely, a notch filter) driven by a uniform clock. Let $H(e^{j\omega T}) = H(e^{j\omega/F})$ be its transfer function. This function can be scaled simply by changing the sampling rate F, resulting in a shift of the filter centre frequency that is proportional to the change of the sampling rate. This simple and natural method is, however, coarse in practice and therefore is usually not suitable for fine tuning of the filter frequency response. This is because the sampling clock frequency F is normally obtained from a fixed high frequency system clock F_c through its division by some integer divider M_c. Thus, the smallest obtainable relative changes $\Delta F/F$ in sampling rate F are limited by the smallest possible relative changes of this divider, i.e. by $\pm 1/M_c$. Such precision can be insufficient for fine frequency response correction, owing to capacitor ratio inaccuracies in a monolithic SC filter.

In order to increase the resolution of control of the filter frequency response, say M times, we can change only every Mth sampling period by some amount, e.g. a multiple of the smallest possible change $\pm T_c = \pm 1/F_c$. This results in a particular heteromerous commensurate sampling with elementary period T_c. Individual sampling periods are

$$T_1 = T_2 = \ldots = T_{M-1} = T \text{ and } T_M = T + \Delta T \quad (5.7)$$

where

$$T = M_c T_c \text{ and } \Delta T = m_c T_c .$$

Thus, the overall period T_o of this heteromerous sampling is given by

$$T_o = MT + \Delta T = NT_c \quad (5.8)$$

where

$$N = M \cdot M_c + m_c . \quad (5.9)$$

The average sampling period Ta can be computed as

$$Ta = T_o/M = T + \frac{\Delta T}{M} = \left(1 + \frac{1}{M}\frac{\Delta T}{T}\right)T = \left(1 + \frac{m_c}{MM_c}\right)T . \quad (5.10)$$

If condition

$$\Delta T/T \ll 1 \text{ ,i.e. } M \gg m_c \quad (5.11)$$

is fulfilled, this sampling technique results merely in shift of the filter centre frequency. The relative shift of the centre frequency is equal to the relative change of the sampling rate

$$\frac{\Delta F}{F} = \frac{Fa - F}{F} = -\frac{\frac{1}{M}\frac{\Delta T}{T}}{1 + \frac{1}{M}\frac{\Delta T}{T}} \approx -\frac{1}{M}\frac{\Delta T}{T} = -\frac{m_c}{MM_c} . \quad (5.12)$$

The above analysis is simplified. The exact analysis should be based on the theory presented in sections 4.2, 4.3 and 4.4. Equation (4.15a) shows that the shift of the filter centre frequency is not the only effect of the considered technique. An undesirable effect also occurs that consists in the appearance of sidebands around the fundamental band, spaced from it every

$$\Delta f = F_\text{o} = NF_\text{c}$$

where $F_\text{o} = 1/T_\text{o}$ and $F_\text{c} = 1/T_\text{c}$. The magnitude of these sidebands is, however, negligible if condition (5.11) is fulfilled.

6

Analysis of multiphase switched-capacitor networks

Multirate and nonuniformly (e.g. heteromerously and commensurately) sampled switched-capacitor networks are inherently multiphase, i.e. driven by a multiphase clock (cf. Fig. 2.5). Among many different approaches to the analysis of multiphase SC networks, two main categories of methods can be distinguished:

- numerical methods (e.g. Vandewalle, De Man and Rabaey, 1981; Grimbleby, 1990)

- symbolic methods (Bon and Kończykowska, 1981; Bedrosian and Refai, 1983; Moschytz and Brugger, 1984; Moschytz and Mulawka, 1986; Kończykowska and Bon, 1988; Dąbrowski and Moschytz, 1989; Unbehauen and Cichocki, 1989; Lin, 1991; Fino, Franca and Steiger-Garção, 1995a, b).

Most of them are dedicated for computer-aided analysis and design of SC networks (Vlach, Vlach and Singhal, 1984; Suyama and Fang, 1992). We are, however, interested in this chapter in a by-inspection and hand analysis of small and medium-sized multirate/multiphase SC networks, e.g. SC modules and building-blocks of larger systems. Direct circuit analysis in individual switching phases and a direct signal-flow graph approach have proven to be very useful for this purpose (Dąbrowski and Moschytz, 1990a, b).

6.1 CIRCUIT ANALYSIS

The simplest but nevertheless quite powerful method for the by-inspection and hand analysis of small multiphase SC building-blocks

130 *Analysis of multiphase switched-capacitor networks*

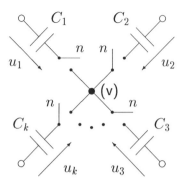

Fig. 6.1 SC circuit node (v) in nth phase.

consists in the classical node analysis of the circuit in individual switching phases using the well-known **charge preservation principle**. Consider, for example, node (v) (Fig. 6.1) just before and just after closing the switches connecting capacitors $C_1, C_2, \ldots, C_k$ to this node in nth phase. Denote voltages across these capacitors as $u_{1a}, u_{2a}, \ldots, u_{ka}$ and $u_{1b}, u_{2b}, \ldots, u_{kb}$ just after and just before closing the switches, respectively. Using the charge preservation principle, we can now write

$$\sum_{\kappa=1}^{k} C_\kappa u_{\kappa a} = \sum_{\kappa=1}^{k} C_\kappa u_{\kappa b} \ . \tag{6.1}$$

Solving such equations for all independent nodes in all switching phases accomplishes the circuit analysis.

We shall present this method on two illustrative examples. Consider first the Nagaraj delay element presented in Fig. 6.2 (Nagaraj, 1984). The main (i.e. working) capacitors are denoted by C_1 and C_2. Additionally, C_{s1}, C_{s2} and C_s are stray capacitances. The effects of op amp DC offset voltage, the DC clock-feedthrough component and switch leakage currents are all modelled by the voltage source u_off connected between the positive op amp input and ground. For the analysis of this circuit, we use the equivalent circuits in individual phases shown in Fig. 6.3. We consider the charge preservation at node (v) in phase 2 only. The analysis in phase 1 is straightforward. Adapting equation (6.1) and denoting the output voltage in

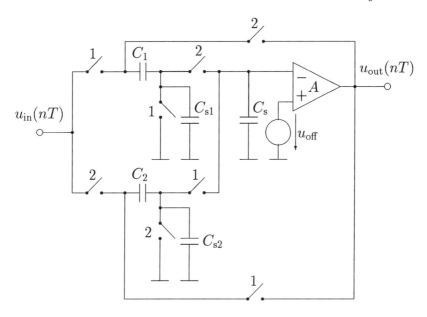

Fig. 6.2 Nagaraj delay element.

phase 2 by $u_{\text{out}}(nT)$ (T being the duration of one clock phase), we get

$$C_1\left(u_{\text{out}}(nT) - u_{\text{off}} + \frac{1}{A}u_{\text{out}}(nT)\right) + (C_{\text{s1}} + C_{\text{s}})\left(-u_{\text{off}} + \frac{1}{A}u_{\text{out}}(nT)\right)$$

$$= C_1 u_{\text{in}}((n-1)T) + C_{\text{s1}} \cdot 0 + C_{\text{s}}\left(-u_{\text{off}} + \frac{1}{A}u_{\text{out}}((n-1)T)\right) \quad (6.2)$$

where A is the DC op amp gain. Equation (6.2) can be rewritten as follows

$$\left(C_1 + \frac{C_1 + C_{\text{s1}} + C_{\text{s}}}{A}\right)u_{\text{out}}(nT)$$

$$= C_1 u_{\text{in}}((n-1)T) + \frac{C_{\text{s}}}{A}u_{\text{out}}((n-1)T) + (C_1 + C_{\text{s1}})u_{\text{off}} . \quad (6.3)$$

Note that this delay element is not offset-compensated and is stray-capacitance sensitive. It is, however, insensitive to capacitor mismatch, since no recharge processes occur.

As the second illustrative example, an SC delay element of Fig. 6.4 is analysed, as proposed in (Dąbrowski, Menzi and Moschytz,

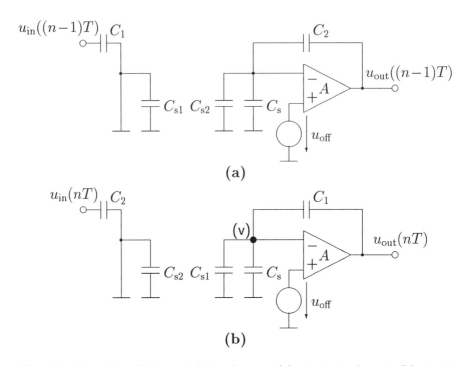

Fig. 6.3 Operation of Nagaraj delay element: (a) circuit in phase 1, (b) circuit in phase 2.

1989). This circuit is offset-compensated and is insensitive to stray capacitance and to capacitor mismatch. Two versions of this element exist (cf. section 7.1.1). One of them, namely the so-called **even delay element**, is shown in Fig. 6.4. To compensate for offset-voltages, a four-phase clock is used. The equivalent circuit in phases 1 and 3 is shown in Fig. 6.5a and that in phases 2 and 4 is shown in Fig. 6.5b. The DC offset-voltage of the op amp is modelled as above by voltage source u_off at the noninverting op amp input. Assuming that the circuit of Fig. 6.5b is depicted at the discrete time instant nT (T being, as before, the duration of one clock phase), we obtain the following relationship between the output voltage $u_\text{out}(nT)$ and the input voltage $u_\text{in}(nT)$ in terms of the finite DC op amp gain A

$$u_\text{out}(nT) - u_\text{off} + \frac{1}{A}u_\text{out}(nT) = u_\text{in}((n-3)T) - \frac{A}{1+A}u_\text{off} \ . \quad (6.4)$$

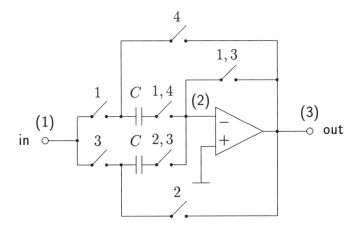

Fig. 6.4 Offset-compensated (even) delay element.

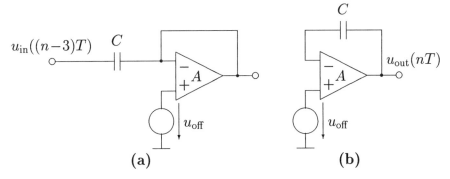

Fig. 6.5 Operation of offset-compensated (even) delay element: (a) circuit in phases 1 and 3, (b) circuit in phases 2 and 4.

After some manipulation, we obtain

$$u_{\text{out}}(nT) = \frac{1}{1+A^{-1}} u_{\text{in}}((n-3)T) + \frac{1}{(1+A^{-1})(1+A)} u_{\text{off}} . \quad (6.5)$$

Since normally $A \geq 5000$, the effects of finite DC gain A in equation (6.5) can be omitted. Thus, we conclude that the DC offset-voltage of the op amp does not affect the output voltage. Moreover, $u_{\text{out}}(nT)$ is not dependent on the value of the capacitors, i.e. it is insensitive to capacitor mismatch. It is also worth noting that the circuit requires a four-phase clock, but only two clock phases per sample.

6.2 SIGNAL-FLOW GRAPH ANALYSIS

Signal-flow graph (SFG) techniques have proven to be very useful for the design and hand analysis of small and medium-sized multiphase switched-capacitor networks, e.g. of SC modules and building-blocks of larger systems (Bedrosian and Refai, 1983; Moschytz and Brugger, 1984; Dostal and Mikula, 1985; Zhou and Shen, 1985; Moschytz and Mulawka, 1986; Dąbrowski and Moschytz, 1989, 1990a; Unbehauen and Cichocki, 1989; Dostál, 1994; Fino, Franca and Steiger-Garção, 1995a, b). Although different methods for the derivation of SFGs for SC networks have been proposed, many of them are still relatively complicated and time consuming. Some of them are also subject to restrictions on the class of networks that can be analysed. Furthermore, most of the methods are not direct, in the sense that they require some intermediate steps comprising auxiliary circuit transformations (Unbehauen and Cichocki, 1989), or even algebraic or topological manipulations (Bon and Kończykowska, 1981; Bedrosian and Refai, 1983; Dostal and Mikula, 1985; Mulawka and Moschytz, 1985; Zhou and Shen, 1985). The direct methods are limited to the analysis of biphase (Moschytz and Brugger, 1984; Moschytz and Mulawka, 1986) or stray-insensitive (Moschytz and Mulawka, 1986; Dąbrowski and Moschytz, 1989) SC networks.

The by-inspection SFG analysis method described in this section is based on the procedure presented by Dąbrowski and Moschytz (1990a). This method is quite general; no restrictions on the networks are required. It is suitable for the analysis of multirate and multiphase SC networks. It is also very simple and direct, in the sense that it allows an immediate, by-inspection derivation of the SFG merely on the basis of a given SC network. The method is primarily useful for the hand analysis and design of small- and medium-sized SC networks (e.g. building-blocks of modular SC systems). It may, however, also serve as a tool for the symbolic computer analysis and for the synthesis of general multirate and multiphase SC networks.

Since new technologies (such as GaAs) relax the restriction of parasitic insensitivity (while admittedly imposing others) (Harrold, Vance and Haigh, 1985; Toumazou and Haigh, 1987), not only stray-insensitive but also general SC circuits with, say, unity-gain buffer amplifiers instead of conventional op amps may be expected in the near future. This was the motivation for extending the SFG analysis method beyond typical stray-insensitive type SC circuits so as to cover the most general multirate and multiphase SC networks.

6.2.1 Signal-flow graph concepts

A signal-flow graph can be considered as a graphical (topological) interpretation of algebraic equations describing the linear system (Mason, 1953, 1956; Coates, 1959; Desoer, 1960; Bedrosian and Refai, 1983). The SFG approach helps to solve these equations in a systematic symbolic way. For example, the system transfer function can be computed by inspection using the so-called **topological formulae**.

A multiphase SC network can be described by equations of the form of (6.1), but to each relevant network node in each switching phase a single such equation must correspond (Hökenek and Moschytz, 1980). Assume that voltages which occur in the network just after closing the switches controlled by the nth switching phase, say in time t_n, are assigned to a time instant t_n. For example, by $v^{(v)}(t_n)$, voltage at node (v) is designated just after switches that are controlled by nth switching phase close in time instant t_n.

We shall consider a multiphase SC network driven by a symmetrical (uniform) M-phase clock as shown in Fig. 2.5, with pulse-width T corresponding to the operation (or sampling) rate $F = 1/T$. Polyphase components of this network operate with a lower operation (sampling) rate $F_o = F/M$. Generalization of our considerations to a nonsymmetric multiphase clock, i.e. to heteromerous sampling, is straightforward but not relevant to the analysis described here. For commensurate sampling we must simply consider a quicker N-phase clock with pulse-width T_c, where the number N is given by equation (5.3).

The $\mathcal{Z}$ transform of voltage $v^{(\nu)}(t_n)$ can be expressed as

$$V^{(\nu)}(z) = \sum_{\nu=0}^{M-1} V_\nu^{(\nu)}(z) \qquad (6.6)$$

where variable z is defined with respect to period T and signals $V_\nu^{(\nu)}(z)$ correspond to polyphase components according to equation (3.53a)*. Let $\mathbf{V}^{(\nu)}(z)$ be a vector composed of signals $V_\nu^{(\nu)}(z)$, i.e.

$$\mathbf{V}^{(\nu)}(z) = \begin{bmatrix} V_0^{(\nu)}(z) \\ V_1^{(\nu)}(z) \\ \vdots \\ V_{M-1}^{(\nu)}(z) \end{bmatrix}. \qquad (6.7)$$

Forming finally the node voltage vector

$$\mathbf{V}(z) = \begin{bmatrix} \mathbf{V}^{(1)}(z) \\ \mathbf{V}^{(2)}(z) \\ \vdots \\ \mathbf{V}^{(k)}(z) \end{bmatrix} \qquad (6.8)$$

by concatenation of vectors (6.7) for all relevant (independent) nodes in the SC network under consideration, say for nodes (1), (2), ..., (k), the charge preservation principle can be written in a form of the following matrix equation

$$\mathbf{C}(z)\mathbf{V}(z) = \mathbf{Q}(z) \qquad (6.9)$$

where $\mathbf{Q}(z)$ is a charge injection (excitation) vector (representing charge injected to each node in each switching phase by energy sources), $\mathbf{V}(z)$ is the node voltage vector as just explained, and $\mathbf{C}(z)$ is a capacitance matrix constructed with respect to constraints introduced by active elements (e.g. op amps).

The idea of SFG construction consists in assigning a graph node to each element of vectors $\mathbf{Q}(z)$ and $\mathbf{V}(z)$. For example, the SFG

---

*Note that we now use variable z as the argument of functions $V_\nu^{(\nu)}(z)$, $\nu = 0, 1, \ldots, M-1$, instead of variable z^M in the polyphase components of equation (3.53a). As a result, our functions $V_\nu^{(\nu)}(z)$ are now analytic rational functions of variable z, which is relevant for the SFG derivation method described below.

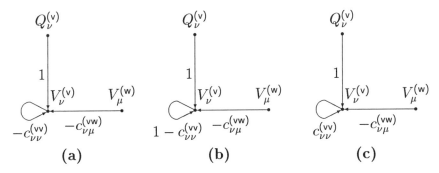

Fig. 6.6 Illustration of different SFG concepts: (a) Coates approach, (b) Mason approach, (c) Bedrosian and Refai approach.

nodes designated as $Q_\nu^{(v)}$ and $V_\nu^{(v)}$ represent polyphase components $Q_\nu^{(v)}(z)$ and $V_\nu^{(v)}(z)$ of signals $\mathbf{Q}^{(v)}(z)$ and $\mathbf{V}^{(v)}(z)$, respectively*. In other words, signal

$$q_\nu^{(v)}(t_n) = \mathcal{Z}^{-1}\{Q_\nu^{(v)}\}$$

is composed of charges injected to node (v) in every νth switching phase, while signal

$$v_\nu^{(v)}(t_n) = \mathcal{Z}^{-1}\{V_\nu^{(v)}\}$$

represents voltages between node (v) and ground in every νth switching phase.

If we rewrite matrix equation (6.9) as a set of scalar equations, then the graph node $V_\nu^{(v)}$ represents the νth equation for network node (v).

The SFG branches represent, in turn, switched capacitors, or more precisely, charges transferred by capacitors between particular nodes in particular switching phases.

There exist three, only slightly different concepts for the construction of SFGs: the Coates approach, the Mason approach and the Bedrosian-Refai approach (Fig. 6.6).

Coates graphs are constructed on the basis of equation (6.9) rewritten in the form

$$\mathbf{Q}(z) - \mathbf{C}(z)\mathbf{V}(z) = \mathbf{0} \ .$$

---

*Variable z is omitted in the SFG node and branch notations for simplicity.

This approach is illustrated by a simple subgraph in Fig. 6.6a, which presents the relations between the three nodes $Q_\nu^{(v)}$, $V_\nu^{(v)}$ and $V_\mu^{(w)}$ of the Coates graph (nodes $V_\nu^{(v)}$ and $V_\mu^{(w)}$ being chosen arbitrarily). Branch gains in Fig. 6.6a, which are denoted without indicating their dependence on variable z for simplicity, are equal to appropriate elements of matrix $\mathbf{C}(z)$ but with opposite sign, namely to elements $-c_{\nu\nu}^{(vv)}(z)$ and $-c_{\nu\mu}^{(vw)}(z)$.

A Mason SFG is constructed from equation (6.9) after being rearranged as

$$\mathbf{Q}(z) + [\mathbf{U} - \mathbf{C}(z)]\mathbf{V}(z) = \mathbf{V}(z)$$

where a unity matrix is denoted by $\mathbf{U}$. The resulting subgraph, again for the nodes $Q_\nu^{(v)}$, $V_\nu^{(v)}$ and $V_\mu^{(w)}$, is shown in Fig. 6.6b. The only formal difference compared with the previous Coates subgraph is that the self-loop gain is greater by 1. There exists, however, a great conceptual difference between the Mason and Coates approaches. Not only do the nodes of a Mason SFG represent individual signals and their respective equations, but the signal values are really constructed in the nodes. Thus, Mason graphs are signal-flow graphs in the strict sense, while Coates graphs are not. Therefore, for Coates SFGs a shorter name is often used, namely **flow graphs** with the word 'signal' omitted (Desoer, 1960). However, in this book we use the notion **signal-flow graph** (SFG) for various existing approaches. This conceptual superiority of Mason SFGs is the reason for their popularity and importance compared to other approaches. Therefore, Mason SFGs are primarily used in this book.

A change of signs of the Coates graph self-loop gains was proposed by Bedrosian and Refai (1983) and resulted in a new SFG concept. The illustrative subgraph is presented in Fig. 6.6c. Now the gains of the self-loops are equal to the corresponding elements of matrix $\mathbf{C}(z)$.

Rearranging equation (6.9) again to the form

$$\mathbf{Q}(z) - \mathbf{C}(z)\mathbf{V}(z) = \mathbf{Q}_0(z) \equiv \mathbf{0}$$

but now using the Mason approach, results in a modified Mason

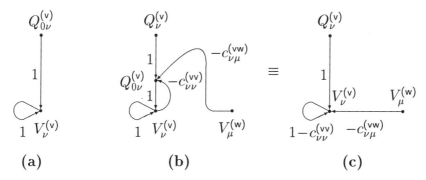

Fig. 6.7 Illustration of modified Mason SFG derivation: (a) subgraph with infinite gain, (b) modified Mason subgraph, (c) equivalent Mason subgraph.

graph, as illustrated in Fig. 6.7. In order to fulfil condition

$$\mathbf{Q}_0(z) \equiv \mathbf{0}$$

we must guarantee that all signals

$$Q_{0\nu}^{(v)}(z) \equiv 0 \ .$$

Therefore, we connect each node $Q_{0\nu}^{(v)}$ to an arbitrarily chosen independent SFG node by a subgraph of infinite gain. For example, we connect node $Q_{0\nu}^{(v)}$ to node $V_{\nu}^{(v)}$ by a subgraph of Fig. 6.7a. The resulting subgraph is presented in Fig. 6.7b. It can be by inspection rearranged to the typical Mason subgraph (cf. Fig. 6.7c).

Having got any variant of the SFG for a given system – the Coates SFG, the Mason SFG or the Bedrosian-Refai SFG – the other two SFG types can immediately be obtained by inspection, simply by appropriate modification of self-loop gains.

6.2.2 Topological formulae

Our purpose is to solve by topological rules the matrix equation (6.9). Usually, most of elements of vector $\mathbf{Q}(z)$ are signals equal to zero. The rest is of secondary importance and we would not like to consider them as input signals but rather as responses to the respective node voltages driven by ideal voltage sources. Let us therefore rearrange equation (6.9) and rename its elements in such a way that the following matrix equation results

$$\mathbf{A}(z)\mathbf{Y}(z) = \mathbf{X}(z) \qquad (6.10)$$

where $\mathbf{X}(z)$ is the excitation vector (i.e. the vector of input signals $X_k(z)$, $k = 1, 2, \ldots, K$), $\mathbf{Y}(z)$ is the response vector (the vector of output signals $Y_k(z)$, $k = 1, 2, \ldots, K$) and $\mathbf{A}(z)$ is the coefficient matrix. Equation (6.10) can also be rewritten as a set of K linear algebraic equations

$$\sum_{l=1}^{K} a_{kl}(z) Y_l(z) = X_k(z) \text{ for } k = 1, 2, \ldots, K . \qquad (6.11)$$

If matrix $\mathbf{A}(z)$ is nonsingular, the solution for signal $Y_l(z)$ is given by

$$Y_l(z) = \frac{\sum_{k=1}^{K} \Delta_{kl}(z) X_k(z)}{\Delta(z)} \text{ for } l = 1, 2, \ldots, K \qquad (6.12)$$

where $\Delta(z)$ is the determinant of the coefficient matrix $\mathbf{A}(z)$, called also the SFG determinant, and $\Delta_{kl}(z)$ is the cofactor of matrix $\mathbf{A}(z)$ of the element at the kth row and lth column.

The transfer function from the SFG node $X_k(z)$ to the node $Y_l(z)$ is thus given by

$$H_{kl}(z) = \frac{Y_l(z)}{X_k(z)} = \frac{\Delta_{kl}(z)}{\Delta(z)} . \qquad (6.13)$$

We shall now examine topological formulae for evaluating determinant $\Delta(z)$ and cofactor $\Delta_{kl}(z)$ simply by inspection from the SFG, but first we have to define some topological notions.

SFG **nodes** represent signals $X_k(z)$ and $Y_k(z)$, $k = 1, 2, \ldots, K$, and **branches** represent elements $a_{kl}(z)$ of matrix $\mathbf{A}(z)$. Nodes corresponding to signals $X_k(z)$ are referred to as the **input nodes**.

A **path** is a succession of consecutive branches, directed in the same way. **Path gain** is the product of individual branch gains forming the path.

A **loop** is a path that starts and ends in the same node but with no other node encountered more than once. The product of the loop branch gains is called **loop gain**. The loops are said to be **joint** if they share at least a common node. Otherwise they are called **disjoint**. A **self-loop** at node (v) is a loop formed by only one branch starting and ending at node (v). If there is no self-loop at node (v), we assume that there exists a self-loop at this node with gain equal to zero. A **composite loop** is any loop which is not a self-loop.

A **connection** is a subgraph of the given SFG such that: each input node is excluded, all other nodes are included, and each included node has only one branch terminating to it and one branch originating from it. **Connection gain** is the product of the branch gains of branches of that connection.

Denote by K_p the number of paths from node $X_k(z)$ to node $Y_l(z)$. Let $P_\kappa(z)$, $\kappa = 1, 2, \ldots, K_p$, be the κth path gain. Topological formulae for evaluating the transfer function $H_{kl}(z)$ have the following general form

$$H_{kl}(z) = \frac{Y_l(z)}{X_k(z)} = \frac{\sum_{\kappa=1}^{K_p} P_\kappa(z) \Delta_\kappa(z)}{\Delta(z)} \qquad (6.14)$$

where $\Delta(z)$ is the SFG determinant and $\Delta_\kappa(z)$ is the determinant of an SFG subgraph that is disjoint from κth path. Thus, the problem of utilization of formula (6.14) reduces to the derivation of an efficient topological formula for evaluating a graph (and a subgraph) determinant.

The most popular is the following well known Mason's formula (Mason, 1956) that refers to Mason SFGs

$$\Delta(z) = 1 - \sum_\kappa L_\kappa(z) + \sum_{\kappa,\lambda} L_\kappa(z) L_\lambda(z) - \sum_{\kappa,\lambda,\mu} L_\kappa(z) L_\lambda(z) L_\mu(z) + \ldots$$
(6.15a)

where the first sum is a sum of all loop gains, the second sum is a sum of the products of the gains of all pairs of disjoint loops, the third sum is a sum of the products of the gains of all triples of disjoint loops, and so on.

The Mason approach is not the optimum, in the sense that cancellation of terms can, in general, occur in the algebraic sums in formula (6.15a). Therefore Coates (1959) and Desoer (1960) proposed an optimum formula, i.e. one with a minimum number of terms and thus no cancellation among them. This formula can be written as

$$\Delta(z) = \sum_{\kappa=1}^{K_c} (-1)^{l_\kappa} C_\kappa(z) \qquad (6.15b)$$

where K_c is the number of all connections in the Coates graph, $C_\kappa(z)$ is the connection gain of the κth connection and l_κ is the number of loops in the κth connection.

Bedrosian and Refai (1983) proposed the following optimum formula

$$\begin{aligned}\Delta(z) &= W(z) - \sum_\kappa T_\kappa^{(1)}(z) W_\kappa^{(1)}(z) + \sum_\kappa T_\kappa^{(2)}(z) W_\kappa^{(2)}(z) \\ &\quad - \sum_\kappa T_\kappa^{(3)}(z) W_\kappa^{(3)}(z) + \ldots \end{aligned} \qquad (6.15c)$$

where $W(z)$ is the product of the gains of all self-loops in Bedrosian–Refai graph, $T_\kappa^{(n)}(z)$ is the product of gains of κth n-tuple of disjoint composite loops and $W_\kappa^{(n)}(z)$ is the product of the gains of those self-loops which are disjoint from the κth n-tuple of composite loops. The summation in the nth sum is taken over all possible such n-tuples. If there are no left nodes, i.e. no self-loops for κth n-tuple, then $W_\kappa^{(n)}(z) \equiv 1$.

We shall illustrate formulae (6.15a)–(6.15c) with a simple example in Fig. 6.8. For the Mason graph in Fig. 6.8b, we get from formula (6.15a)

$$\begin{aligned}\Delta(z) &= 1 - (1 - a_{11}) - (1 - a_{22}) - (-a_{12})(-a_{21}) \\ &\quad + (1 - a_{11})(1 - a_{22}) \\ &= 1 - 1 + a_{11} - 1 + a_{22} - a_{12}a_{21} + 1 - a_{11} - a_{22} + a_{11}a_{22} \\ &= a_{11}a_{22} - a_{12}a_{21}\end{aligned}$$

where the dependence on variable z is omitted for simplicity. The effect of term cancellation is self-explanatory. This effect does not occur in the Coates–Desoer formula (6.15b) for the Coates graph

Signal-flow graph analysis 143

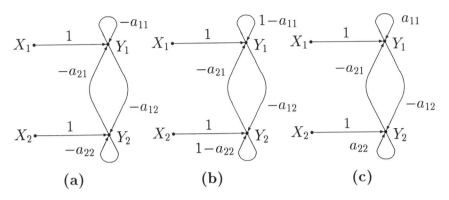

Fig. 6.8 Illustrative example of different SFG concepts: (a) Coates SFG, (b) Mason SFG, (b) Bedrosian–Refai SFG.

in Fig. 6.8a

$$\Delta(z) = (-1)^2(-a_{11})(-a_{22}) + (-1)^1(-a_{12})(-a_{21})$$
$$= a_{11}a_{22} - a_{12}a_{21}$$

and in the Bedrosian–Refai formula (6.15c) for the graph in Fig. 6.8c

$$\Delta(z) = a_{11}a_{22} - (-a_{12})(-a_{21}) \cdot 1$$
$$= a_{11}a_{22} - a_{12}a_{21}.$$

Using formula (6.14), we can now determine the following transfer functions

$$H_{11}(z) = \frac{Y_1(z)}{X_1(z)} = \frac{a_{22}}{a_{11}a_{22} - a_{12}a_{21}},$$

$$H_{12}(z) = \frac{Y_2(z)}{X_1(z)} = \frac{-a_{12}}{a_{11}a_{22} - a_{12}a_{21}},$$

$$H_{21}(z) = \frac{Y_1(z)}{X_2(z)} = \frac{-a_{21}}{a_{11}a_{22} - a_{12}a_{21}}$$

and

$$H_{22}(z) = \frac{Y_2(z)}{X_2(z)} = \frac{a_{11}}{a_{11}a_{22} - a_{12}a_{21}}.$$

6.2.3 SSN-type SC networks

In this section we assume that the only active elements contained in SC networks under consideration are ideal inverting op amps and possibly also finite-gain amplifiers controlled by op amp outputs. Such SC networks contain only three types of nodes (except for the grounded reference node):

- nodes driven by ideal voltage sources (e.g. op amp outputs), the **source nodes**

- nodes at virtual ground, the **sink nodes**

- nodes connecting capacitors with switches; these nodes do not appear in the SFG and need not be considered further because they are also connected either to the source or to sink nodes or are disconnected from the network (when corresponding switches are open).

The op amp outputs are referred to as the **independent** source nodes and the finite-gain amplifier outputs as the **dependent** source nodes. Note that an independent source node can be represented by a grounded norator, and a sink node by a grounded nullator (Moschytz, 1987). Thus, the number of independent source nodes must always be equal to the number of sink nodes, unless the network is ill-conditioned.

We refer to SC networks belonging to the class defined above as **source-sink-node-type** (SSN) SC networks[*]. All possible switching transitions for the capacitor plate of an SSN-type SC network are illustrated in Fig. 6.9. Switch connections designated by dotted lines are not allowed for stray-insensitive SC networks (Hasler, 1981). Hence, the class of SSN-type SC networks is slightly broader than that of stray-insensitive networks[†].

In the following considerations we shall assume that the only source nodes in an SSN-type SC network are independent source

---

[*]Note that the definition of SSN-type networks can be applied to any active network; SSN-type networks are those containing source and sink nodes only.

[†]General SC networks, i.e. those which do not belong to the class of SSN-type SC networks, cannot be stray-insensitive. Nevertheless, they can be stray-compensated (cf. Figs 7.5d, 7.7 and equation (7.4)).

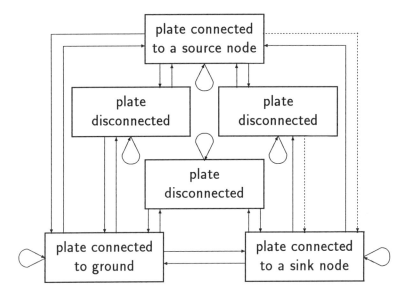

Fig. 6.9 Switching possibilities for any capacitor plate of a multiphase SSN-type SC network.

nodes, i.e. that the only active elements contained in such a network are inverting op amps. This is a minor restriction and its generalization is straightforward.

From considerations in section 6.2.1 (cf. Fig. 6.7), we conclude that the SFG of an SSN-type SC network containing only ideal inverting op amps is composed of two types of subgraphs, namely:

- subgraphs modelling op amps (Fig. 6.10a)

- branches modelling switched capacitors (Fig. 6.10b).

Op amps are modelled by subgraphs with infinite gain in Fig. 6.7a. Thus, the resulting SFG will contain only source and sink nodes*; these correspond to the source and sink nodes of the SC network, split, however, into individual switching phases. More precisely, the SFG nodes correspond to the op amp output voltages $V^{(v)}$ for the source nodes and to the op amp input zero-valued charges $Q_0^{(v)}$ for

---

*Note that all SFG sink nodes may be eliminated by tying them to the corresponding source nodes as shown in Fig. 6.7b and c.

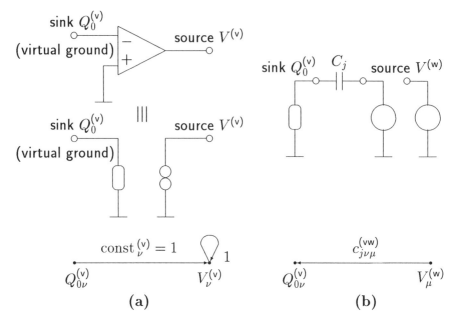

Fig. 6.10 SFG subgraphs: (a) subgraph modelling (v)th op amp in νth switching phase, (b) branch modelling capacitor C_j switched between (w)th source node in μth phase and (v)th sink node in νth phase.

the sink nodes, decomposed into individual polyphase components, i.e. to voltages $V_\nu^{(v)}$ and charges $Q_{0\nu}^{(v)}$, respectively, for the (v)th op amp and for individual switching phases $\nu = 0, 1, \ldots, M-1$. Thus, each source and each sink node of the M-phase SC network generates at most M nodes in the SFG, and each op amp at most M subgraphs of the type in Fig. 6.10a. If for any switching phase, e.g. for the phase ζ, the inverting input and the output of the (v)th op amp are short circuited by a switch closed in that phase, then the nodes $V_\zeta^{(v)}$ (source node) and $Q_{0\zeta}^{(v)}$ (sink node), and the corresponding subgraph of the type in Fig. 6.10a, do not appear in the SFG.

Note that op amp subgraphs (Fig. 6.10a) always start at sink nodes and end in source nodes, whereas the capacitor branches start at source nodes and end in sink nodes. The transfer function designated in Fig. 6.10a as $\mathrm{const}_\nu^{(v)}$ is an arbitrary number but not equal to zero. Therefore, choosing the simplest and most convenient

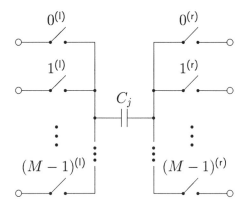

Fig. 6.11 A multiphase switched capacitor C_j.

possibility, we let $\mathrm{const}\,{}_\nu^{(\mathbf{v})} = 1$ for all ν and $(\mathbf{v})$ (cf. Fig. 6.7a). With this assumption the only parameters appearing in the SFG are the transfer functions $c_{j\nu\mu}^{(\mathrm{vw})}$, where μ is the switching phase at the source node $V^{(\mathrm{w})}$ and ν the switching phase at the sink node $Q_0^{(\mathrm{v})}$ (Fig. 6.10b). We shall now show that these transfer functions can be determined directly by inspection of the SC network. To this end we consider a jth generally multiphase-switched capacitor C_j, as shown in Fig. 6.11. Switches $\nu^{(l)}$ and $\nu^{(r)}$, $\nu = 0, 1, \ldots, M-1$, are closed in the νth phase, provided they are activated*.

The following observations can now be made:

- a capacitor is charged in the μth phase if one of its plates is connected to a source node and the other, directly or indirectly (i.e. via a source node or a sink node), to ground (cf. Fig. 6.12a)

- the charge of a capacitor (acquired during the μth phase) is transferred to a sink node during the νth phase if at least one capacitor plate is connected to a sink node and the other – directly or indirectly – to ground during the νth phase (Fig. 6.12c). Note that in general

$$\nu = (\mu + \xi) \bmod M \text{ for } \xi = 0, 1, \ldots, M-1$$

---

*Switches that are not activated remain open in our model and are omitted in the actual circuit.

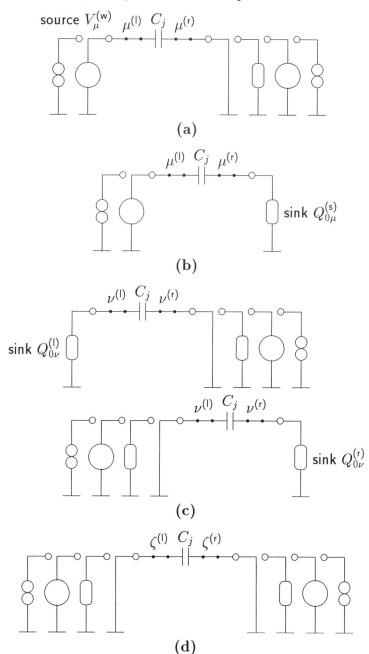

Fig. 6.12 Particular switching situations for a capacitor C_j: (a) charge (μth switching phase), (b) charge transfer (μth switching phase), (c) charge transfer (νth switching phase), (d) discharge (ζth switching phase).

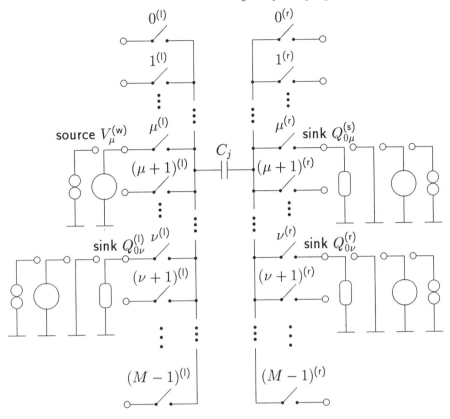

Fig. 6.13 A multiphase switched capacitor C_j of an SSN-type SC network.

- a capacitor is discharged during any phase, say during the phase ζ, $\zeta = 0, 1, \ldots, M-1$, if both capacitor plates are shorted to ground, directly or indirectly, during that phase (Fig. 6.12d).

Note that the conditions for capacitor charge, charge transfer, and discharge are not mutually exclusive; they can take place simultaneously during the same phase (cf. Figs 6.12a and b).

The following observations are direct consequences of those above. In order for a capacitor of an SSN-type SC network to influence the surrounding network, it must have at least one source node and one sink node connected to it in at least one switching phase; without at least one source node at some switching phase, the capacitor will never be charged; on the other hand, without at least one sink

150   Analysis of multiphase switched-capacitor networks

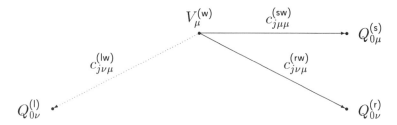

Fig. 6.14 Branches possibly generated in the SFG of an SSN-type SC network by a capacitor C_j charged in the μth phase.

node at some switching phase, the capacitor will never transfer its charge to the surrounding network. Moreover, in some switching phase, both capacitor plates must be connected directly, or indirectly (via a source node or a sink node), to ground in that phase. Otherwise no charge or discharge of that capacitor can take place, and thus, that capacitor will have no influence on the network and can be omitted.

In consequence, a capacitor charged in the μth phase, say by the (w)th source node connected during this phase to the left capacitor plate (Fig. 6.13), can transfer its charge in an SSN-type SC network through at most three sink nodes, designated in Fig. 6.13 as nodes (s), (l) and (r). The (s) node is that connected to the right capacitor plate during the μth phase. Thus, the charge transfer is simultaneous, i.e. it is present in the same phase as the charge of this capacitor. The (l)th node and the (r)th node are connected to the left and to the right capacitor plate, respectively, during the νth phase, ν being any phase within the following time period $T_o = MT$. The charge transfer in the νth phase takes place only if there is no phase ζ in between phases μ and ν in which the capacitor is discharged.

From the above observations we conclude that in the SFG of a multiphase SSN-type SC network, at most three branches, corresponding to a particular capacitor C_j, can leave a source node (Fig. 6.14). Note that the transfer function $c_{j\nu\mu}^{(lw)}$ of the branch, indicated in Fig. 6.14 by a dotted line, results from switching the left plate of the capacitor C_j between a source node and a sink

Table 6.1 Transfer functions of SFG branches resulting from the capacitor C_j connected to source node (w) in phase μ

	Switching configuration		Transfer function value			
Switch $\mu^{(r)}$	Switches $\zeta^{(l)}$ & $\zeta^{(r)}$ ζ is a phase in between phases μ and ν	Switches $\nu^{(l)}$ & $\nu^{(r)}$ ν is a charge transfer phase $\nu = (\mu+\xi)\mathrm{mod}\,M$	Arbitrary SC networks			
			SSN-type SC networks	Stray-insensitive SC networks		
			$c_{j\mu\mu}^{(\mathrm{sw})}$	$c_{j\nu\mu}^{(\mathrm{rw})}$	$c_{j\nu\mu}^{(\mathrm{lw})}$	$c_{j\mu\mu}^{(\mathrm{vw})}$
open	arbitrary	arbitrary	0	0	0	0
closed to a sink node	both closed	arbitrary	C_j			$-C_j$
	at least one open	at least one open	$C_j(1-z^{-M})$			$-C_j(1-z^{-M})$
closed to ground or a source	both closed	arbitrary	0			$-C_j$
	at least one open	at least one open	0			$-C_j(1-z^{-M})$
closed	closed	arbitrary		0	0	
	at least one open	$\nu^{(l)}$ closed $\nu^{(r)}$ closed to a sink node		$-C_j z^{-\xi}$		
		$\nu^{(r)}$ closed $\nu^{(l)}$ closed to a sink node			$C_j z^{-\xi}$	

node. This is, however, not allowed for stray-insensitive networks (cf. Fig. 6.9). Thus, in the SFG of a multiphase stray-insensitive SC network, at most two branches, corresponding to a particular capacitor, can leave a source node. Finally, the values of the transfer functions $c_{j\mu\mu}^{(\mathrm{sw})}$, $c_{j\nu\mu}^{(\mathrm{rw})}$ and $c_{j\nu\mu}^{(\mathrm{lw})}$ are summarized in Table 6.1. They follow directly from above observations and from the following considerations.

Assume, as in Fig. 6.13, that node $\mu^{(l)}$ connects a capacitor C_j to a (w)th source node (then all other nodes in Fig. 6.13 are potential sink nodes).

First, assume that switch $\mu^{(r)}$ connects the capacitor C_j to a sink node (s) during the μth phase. If within the following time period $T_o = MT$ there exists a phase ζ such that both switches $\zeta^{(l)}$ and $\zeta^{(r)}$ are closed, thereby discharging capacitor C_j in phase ζ, then

$$c_{j\mu\mu}^{(\text{sw})} = C_j \tag{6.16a}$$

otherwise, i.e. if no such discharging in any phase ζ occurs

$$c_{j\mu\mu}^{(\text{sw})} = C_j(1 - z^{-M}) \ . \tag{6.16b}$$

Second, assume that switch $\nu^{(r)}$ connects a capacitor C_j to a sink node during a phase $\nu \neq \mu$, thus

$$\nu = (\mu + \xi) \bmod M \text{ for } \xi = 1, \ldots, M - 1 \ . \tag{6.17}$$

If switches $\mu^{(r)}$ and $\nu^{(l)}$ are both closed in their respective phases, and if in no phase ζ between phases μ and ν, i.e.

$$\zeta = (\mu + 1) \bmod M, (\mu + 2) \bmod M, \ldots, (\nu - 1) \bmod M$$

are switches $\zeta^{(l)}$ and $\zeta^{(r)}$ both simultaneously closed, then

$$c_{j\nu\mu}^{(\text{rw})} = -C_j z^{-\xi} \tag{6.18a}$$

otherwise, i.e. if at least one of the switches $\mu^{(r)}$ or $\nu^{(l)}$ is not closed, or if there exists a phase ζ between phases μ and ν such that the switches $\zeta^{(l)}$ and $\zeta^{(r)}$ are both closed (thereby discharging capacitor C_j during the ζth phase, i.e. before the νth phase), then

$$c_{j\nu\mu}^{(\text{rw})} = 0 \ . \tag{6.18b}$$

Finally, assume that switch $\nu^{(l)}$ connects a capacitor C_j to a sink node during a phase ν fulfilling condition (6.17). If switches $\mu^{(r)}$ and $\nu^{(r)}$ are both closed in their respective phases, and if in no phase ζ between phases μ and ν are switches $\zeta^{(l)}$ and $\zeta^{(r)}$ both simultaneously closed, then

$$c_{j\nu\mu}^{(\text{lw})} = C_j z^{-\xi} \tag{6.19a}$$

otherwise, i.e. if at least one of the switches $\mu^{(r)}$ or $\nu^{(r)}$ is not closed, or if there exists a phase ζ between phases μ and ν such that the switches $\zeta^{(l)}$ and $\zeta^{(r)}$ are both closed (thereby discharging capacitor C_j before the νth phase in phase ζ), then

$$c_{j\nu\mu}^{(\text{lw})} = 0 \ . \tag{6.19b}$$

The steps required for the derivation of the SFG of an M-phase SSN-type SC network can be summarized as follows.

- Number all nodes of the SC network and check whether the network belongs to the class of SSN-type networks, i.e. if every node is either a source or a sink node.

- Draw all SFG nodes. First, M SFG nodes (corresponding to M polyphase components) for each network node may be drawn. Then, the nodes disconnected from the network in individual switching phases must be omitted from the SFG. Furthermore, those source–sink-node pairs corresponding to op amps with the inverting inputs shorted to their outputs in individual phases must also be omitted.

- Connect the SFG nodes using op amp subgraphs (subgraphs of infinite gain) of the type presented in Fig. 6.8a and switched capacitor branches as shown in Fig. 6.8b. Transfer functions of the latter are determined on the basis of Table 6.1 or, equivalently, by expressions (6.16a)–(6.19b).

- Finally, simplify the SFG by eliminating all sink nodes by tying them to the corresponding source nodes using the SFG equivalence illustrated in Fig. 6.7b and c. With a little practice, this simplified SFG can be obtained directly by inspection.

The following example illustrates the above procedure. Consider the four-phase offset-compensated delay element of Fig. 6.4. This circuit was proposed by Dąbrowski, Menzi and Moschytz (1989). It contains two source nodes (nodes **(1)** (in) and **(3)** (out) in Fig. 6.4) and one sink node (node **(2)**). Because it is customary in the literature, we number the switching phases 1, 2, 3, 4 (and not 0, 1, 2, 3 as was done, for convenience, in the above theoretical considerations). The first step for the derivation of the SFG of this circuit consists of

154 *Analysis of multiphase switched-capacitor networks*

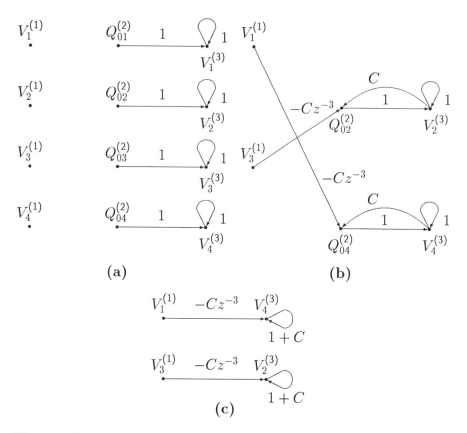

Fig. 6.15 Derivation of SFG for four-phase offset-compensated delay element of Fig. 6.4: (a) potential SFG nodes and op amp subgraphs, (b) resulting SFG, (c) simplified SFG.

drawing all potential SFG nodes and subgraphs modelling op amps. This is illustrated in Fig. 6.15a. SFG nodes $V_1^{(1)}$, $V_2^{(1)}$, $V_3^{(1)}$, $V_4^{(1)}$ and $V_1^{(3)}$, $V_2^{(3)}$, $V_3^{(3)}$, $V_4^{(3)}$ represent circuit source nodes (1) and (3) in switching phases 1, 2, 3, 4, respectively. On the other hand, SFG nodes $Q_{01}^{(2)}$, $Q_{02}^{(2)}$, $Q_{03}^{(2)}$, $Q_{04}^{(2)}$ represent circuit sink node (2). Notice, however, that node (1) is switched off from the circuit in phases 1 and 3. Therefore, nodes $V_1^{(1)}$ and $V_3^{(1)}$ must be withdrawn from the SFG (Fig. 6.15b). Note, moreover, that the op amp should be represented by only two subgraphs of the type in Fig. 6.10a, namely by those corresponding to switching phases 2 and 4. This is because in phases 1 and 3, the inverting input and the output of the op amp are short-circuited by respective switches. In other words, $V_1^{(3)} = V_3^{(3)} \equiv 0$, and therefore these signals are omitted. The left op amp subgraphs connect nodes $Q_{02}^{(2)}$ and $Q_{04}^{(2)}$ with nodes $V_2^{(3)}$ and $V_4^{(3)}$, respectively (Fig. 6.15b). The upper branch with capacitor C generates two branches in the SFG. One of them is that from node $V_1^{(1)}$ to node $Q_{04}^{(2)}$ since the upper capacitor C is connected (i.e. charged) to the source node (1) in phase 1 and to the sink node (2) (i.e. discharged) in phase 4. The corresponding transfer function $-Cz^{-3}$ is given by expression (6.18a) with $\mu = 1$, $\nu = 4$ and $\xi = (4-1) \bmod 4 = 4 - 1 = 3$. The second branch, which also relates to upper capacitor C, starts in node $V_4^{(3)}$ and ends in node $Q_{04}^{(2)}$. Its transfer function C follows from equation (6.16a). This is due to the fact that the upper capacitor C is discharged from the charge corresponding to voltage $V_4^{(3)}$ in phase 1. Similarly, we also derive two branches and their transfer functions, relating to the lower capacitor C. One of them connects node $V_3^{(1)}$ with node $Q_{02}^{(2)}$. Its transfer function $-Cz^{-3}$ follows from formula (6.18a) for $\mu = 3$, $\nu = 2$ and $\xi = (2-3) \bmod 4 = 3$. The other branch, generated by the lower capacitor C, is that from node $V_2^{(3)}$ to node $Q_{02}^{(2)}$ with transfer function C following from equation (6.16a). Finally, tying the sink nodes, which represent the signals $Q_{02}^{(2)}$ and $Q_{04}^{(2)}$, to the corresponding source nodes (i.e. to those with signals $V_2^{(3)}$ and $V_4^{(3)}$, respectively) results in the simplified SFG shown in Fig. 6.15c. This SFG comprises only source nodes. Note that in our example it is composed of two independent subgraphs. One of them starts in node $V_1^{(1)}$ and ends in node $V_4^{(3)}$. The second starts in node $V_3^{(1)}$ and ends in node $V_2^{(3)}$. Thus, applying Mason's topological formula (section 6.2.2), we obtain two possible transfer functions of the network $V_4^{(3)}/V_1^{(1)}$ and $V_2^{(3)}/V_3^{(1)}$, which, as a matter of fact, are equal to each other

$$H(z) = \frac{U_{\text{out}}(z)}{U_{\text{in}}(z)} = \frac{V_4^{(3)}}{V_1^{(1)}} = \frac{V_2^{(3)}}{V_3^{(1)}} = z^{-3} \ .$$

156 *Analysis of multiphase switched-capacitor networks*

Note that above result is consistent with equation (6.5) for op amp gain $A = \infty$ and offset voltage $u_{\text{off}} = 0$.

6.2.4 SSN-network transformation

Although the class of SSN-type SC circuits is slightly broader than that of stray-insensitive networks, it is still quite restrictive, e.g. SC circuits with unity-gain buffer amplifiers (instead of inverting op amps), which are conceivable with advancing new technologies (such as GaAs) do not belong to this class. Therefore, in this section, we generalize the method described in section 6.2.3 in order to be able to cope with multirate and multiphase SC networks with arbitrary structure.

The general method is composed of two steps:

- first, a given arbitrary SC network is transformed into the equivalent SSN-type network by the so-called SSN (**source–sink-node**)**-network transformation** (Dąbrowski and Moschytz, 1990b)

- second, rules presented in section 6.2.3 are applied to derive the SFG by inspection.

Notice that in addition to elements contained in SSN-type networks, general networks may contain:

- **general nodes**, i.e. nodes which are neither source nor sink nodes

- differential input op amps

- finite-gain amplifiers (including unity-gain buffers).

In order to present the SSN-network transformation concept, consider the simple network shown in Fig. 6.16*. It consists of a switched capacitor C_j connected in phase ν between a source node (s) (with node voltage $V_\mu^{(s)}(z)$) and a general node (g) (with node voltage $V_\mu^{(g)}(z)$). Our objective is to transform this network so that

---

*It should be stressed that the SSN-network transformation is not restricted to SC networks and can be applied to any passive or active network.

the general node (g) is broken into two nodes: a source node (w), with the same node voltage as that of the general node, i.e.

$$V_\mu^{(w)}(z) = V_\mu^{(g)}(z),$$

and a sink node (v) modelling the same charge flow but now under the virtual ground circumstances. The transformed (SSN-type) network is equivalent to the original, even though the charge leaving the source node (s) (i.e. $Q_\mu^{(s)}(z) = -Q_\mu^{(g)}(z)$ in Fig. 6.16a) and the charge $Q'^{(s)}_\mu(z)$ in Fig. 6.16d may be different for the two networks (since charges $Q_\mu^{(s)}(z)$ and $Q'^{(s)}_\mu(z)$ are supplied from source nodes, they need not be the same).

To transform the network of Fig. 6.16a into an SSN-type network we connect a series nullator–norator between the general node (g) and ground (Fig. 6.16b). This does not affect the original circuit because a nullator and norator in series correspond to an open circuit (Moschytz, 1974). Nevertheless, it can be interpreted as the insertion of a unity-gain amplifier (gain $\beta = 1$ indicated on the right-hand side of Fig. 6.16b). In order to split the general node (g) into a source and a sink node, we require a grounded norator and nullator, as pointed out earlier. We can achieve this by applying the so-called **complementary transformation** (Fliege, 1973) to the network of Fig. 6.16b, as shown in Fig. 6.16c. However, according to the complementary transformation rules, we now have the new gain $\bar\beta = \beta/(\beta - 1) = \infty$ instead of the unity gain we started out with. There is, however, a second condition for the general node that governs the current charge preservation, and that must also be satisfied, namely

$$Q_\mu^{(s)} = -Q_\mu^{(g)} = -c_{\mu\mu}^{(sg)}(V_\mu^{(g)} - V_\mu^{(s)}) = c_{\mu\mu}^{(sg)}(V_\mu^{(s)} - V_\mu^{(w)})$$
$$= c_{\mu\mu}^{(sg)} V_\mu^{(s)} + c_{\mu\mu}^{(vw)} V_\mu^{(w)}$$

where variable z is omitted for simplicity and the notation

$$c_{\mu\mu}^{(vw)} = -c_{\mu\mu}^{(sg)}$$

is used. The above condition is fulfilled by connecting in phase μ an additional branch with the transfer function $-c_{\mu\mu}^{(sw)}(z)$ between

158   Analysis of multiphase switched-capacitor networks

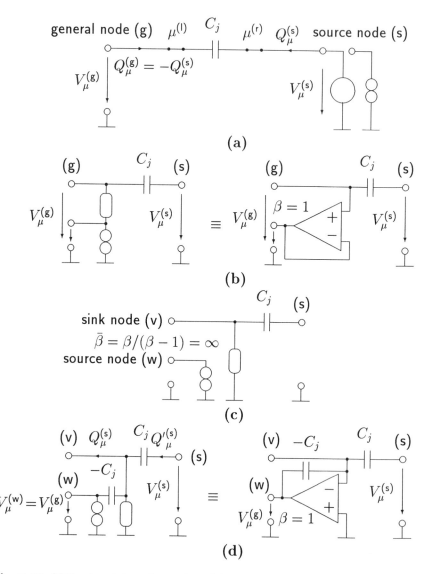

Fig. 6.16 SSN-network transformation: (a) original circuit, (b) equivalent circuit with dummy unity-gain amplifier, (c) complementary transformed circuit, (d) final transformed circuit.

the new source node (w) (grounded norator) and the sink node (v) (grounded nullator) in Fig. 6.16c, thus connecting a capacitance $-C_j$ between these nodes in phase μ, as shown in Fig. 6.16d. This automatically restores our unity gain condition since for $Q_\mu^{(w)}(z) \equiv 0$ we can write

$$\frac{V_\mu^{(w)}(z)}{V_\mu^{(s)}(z)} = -\frac{-c_{\mu\mu}^{(vw)}}{c_{\mu\mu}^{(vw)}} = 1$$

i.e. $\beta = 1$; at the same time it completes the SSN-network transformation for the illustrative network in Fig. 6.16. In the right-hand side column of Table 6.1, possible values for the transfer function $c_{\mu\mu}^{(vw)}$ are listed.

In the context of the presented SFG analysis method, the concept of a transformed SSN-type SC network has primarily theoretical significance. The occurrence of a negative capacitor is, therefore, of no concern.

The SSN-network transformation can readily be applied to any multirate and multiphase SC network. This is achieved by converting each switched capacitor of the original SC network into its SSN-type counterpart for each switching phase. In Fig. 6.17, each possible configuration for a branch with a capacitor C_j incident to at least one general node in at least one switching phase is shown, together with the corresponding SSN-type counterpart. For instance, a capacitor C_j, connected in the μth switching phase between two general nodes results in four capacitors (two positive and two negative) in the transformed SC network. During this phase, they are connected as shown on the right-hand side of Fig. 6.17a. Note that the nullators and norators must be placed in such a way that a nullator and a norator occur at each end of capacitors C_j and $-C_j$. Thus for example, the two norators cannot be placed at the two upper nodes, and the two nullators at the two lower nodes on the right-hand side of Fig. 6.17a.

Consider now a general node (g) with capacitors $C_1, C_2, \ldots, C_j$ connected to it in some switching phase, as illustrated in Fig. 6.18a. After the SSN-network transformation (Fig. 6.18b), we obtain an equivalent circuit containing two nodes: the source node (w) and the sink node (v), which replace the general node (g) of the original circuit. Although any general SC circuit node should, in principle, be transformed into the two nodes of the corresponding SSN-type circuit, simplifications, analogous to Nathan's rules in matrix

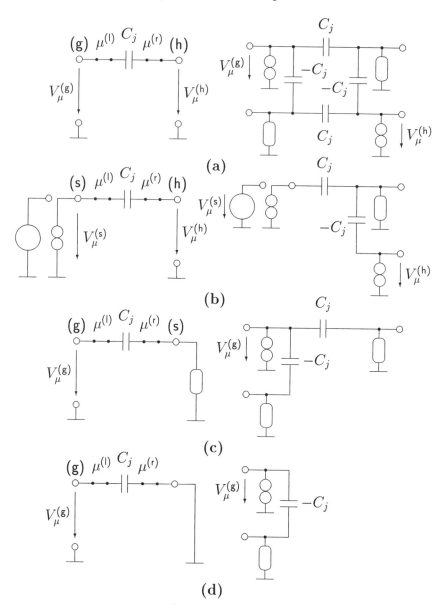

Fig. 6.17 SSN-network transformation of a branch with capacitor C_j in phase μ: (a) branch connected between two general nodes (g) and (h), (b) branch connected between a source node (s) and a general node (h), (c) branch connected between a general node (g) and a sink node (s), (d) branch connected between a general node (g) and ground.

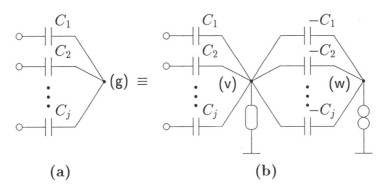

Fig. 6.18 SSN-network transformation of a general node (**g**): (a) initial circuit, (b) final transformed circuit.

active network analysis (Nathan, 1961), can be made, which correspond to constraints imposed by some of the active elements in certain switching phases (Fig. 6.19). Note that the number of SFG sink and independent source nodes are both reduced by the same amount owing to these simplifications, and thus the possibility of node pairing, necessary to connect SFG sink nodes to source nodes by subgraphs of the type in Fig. 6.7a (or Fig. 6.10a), is maintained.

The simplifications mentioned above are as follows.

- The node voltages at the input and output of a unity-gain buffer amplifier are equal; the two nodes are modelled by a single source node representing the common (input/output) voltage and by a single sink node corresponding to the buffer input charge (Fig. 6.19c).

- The two voltages at the input nodes of a differential-input op amp will in general be equal (Fig. 6.19d). Both nodes are combined into a single source node whose voltage corresponds to the common input voltage. The input charges must be modelled separately (i.e. by two different sink nodes), however. The fourth node, necessary for sink–source node pairing, corresponds to the op amp output node.

- The voltages of some nodes can be related to each other, e.g. by finite-gain amplification. In this case only the amplifier inputs

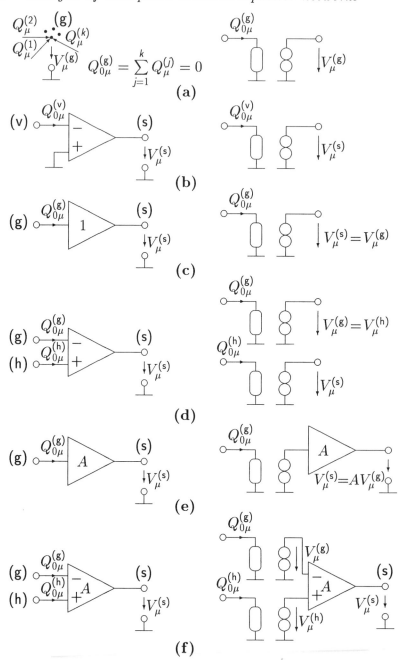

Fig. 6.19 SSN representation of general nodes and active elements: (a) general node, (b) inverting op amp, (c) unity-gain buffer, (d) differential-input op amp, (e) finite-gain differential amplifier, (f) finite-gain amplifier.

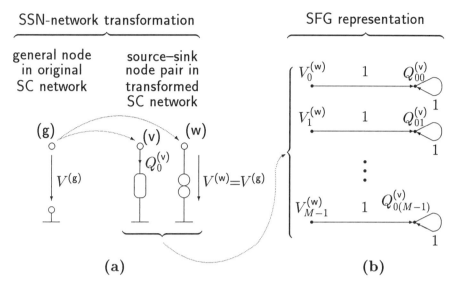

Fig. 6.20 Splitting of a general node (**g**) into the corresponding SFG nodes: (a) SSN-network transformation, (b) SFG representation.

are modelled as general nodes; the outputs are represented in the SFG by additional (dependent) source nodes. These are connected by direct gain branches (gain A) with the SFG source nodes corresponding to the amplifier inputs (Fig. 6.19e and f).

6.2.5 Analysis of general SC networks

In our procedure, the SFG nodes represent the source nodes, sink nodes, and general nodes of the SC network, where the general nodes are split into source and sink nodes as illustrated in Fig. 6.20. Note that a general node, which is transformed into a source–sink node pair, is treated the same as an op amp (compare Figs 6.10a and 6.20). Whereas subgraphs corresponding to active elements are derived by the rules given in section 6.2.4, the branches modelling switched capacitors result from Table 6.1.

Assume that a capacitor C_j of an arbitrary M-phase SC network is connected in the μth phase by at least one of its plates (e.g. by the

164 *Analysis of multiphase switched-capacitor networks*

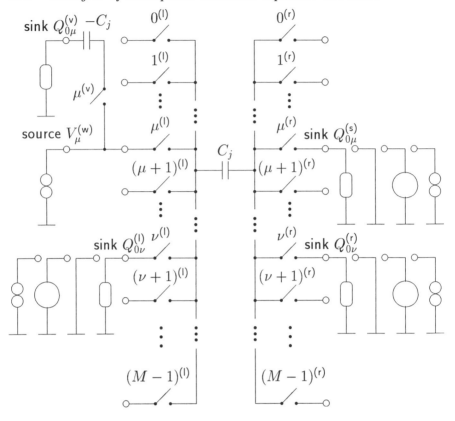

Fig. 6.21 A multiphase switched capacitor C_j of an arbitrary SC network.

left plate) to a general node, say to node **(g)** (Fig. 6.21). In order for any kind of charging or discharging to take place, the right capacitor plate must not be disconnected from the network in this phase. From Fig. 6.17 we conclude that in the transformed SSN-type SC circuit, both the capacitor C_j and the capacitor $-C_j$ are charged to the voltage $V_\mu^{(w)} = V_\mu^{(g)}$ from the source node **(w)**. Indeed, in the μth switching phase, capacitor C_j is connected between the source node **(w)** and the sink node **(v)** (Fig. 6.21), and therefore generates a branch with transfer function $c_{\mu\mu}^{(vw)}$ in the SFG between these two nodes, as shown in Fig. 6.22. Since by definition the capacitor $-C_j$ has the negative charge of the capacitor C_j, the transfer function

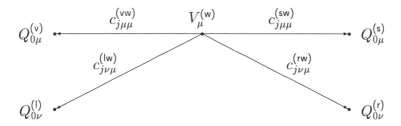

Fig. 6.22 Branches possibly generated in the SFG of an arbitrary SC network by a capacitor C_j charged in the μth phase.

$c_{\mu\mu}^{(vw)}$ of this branch fulfils the following equation:*

$$c_{\mu\mu}^{(vw)} = -c_{\mu\mu}^{(sw)} . \tag{6.20}$$

Note that transfer function $c_{\mu\mu}^{(sw)}$ is determined by expression (6.16a) or (6.16b), depending on the existence of a discharging phase ζ (cf. Table 6.1).

The following observation can now be made. It permits the direct derivation of the SFG from a given general SC network without actually invoking the SSN-network transformation. Let **(g)** be any general node of an arbitrary M-phase SC network. Then, assuming that node **(g)** is treated simultaneously as a source node **(w)** and a sink node **(v)** as illustrated in Fig. 6.20, the transfer functions of the SFG branches that start at source nodes $V_\mu^{(w)}$, $\mu = 0, 1, \ldots, M-1$, and end in sink nodes other than $Q_{0\mu}^{(v)}$, as well as the transfer functions of those branches that start in source nodes other than $V_\mu^{(w)}$ and end in sink nodes $Q_{0\mu}^{(v)}$, $\mu = 0, 1, \ldots, M-1$, are determined by equations (6.16a)–(6.19b). In addition, branches starting at the source node $V_\mu^{(w)}$ and ending in the sink node $Q_{0\mu}^{(v)}$, $\mu = 0, 1, \ldots, M-1$, must be accounted for. These branches correspond to capacitors connected by one of their plates to the node **(g)** in the μth phase, and not disconnected from the network by the other plate in this phase. Let C_j be one such capacitor. The transfer function $c_{\mu\mu}^{(vw)}$

---

*Here we assume, of course, that the other plate of capacitor C_j is connected to the sink node in the μth phase (as shown in Fig. 6.17c) and not to a source node or to ground, in which case the transfer function $c_{\mu\mu}^{(sw)}$ would equal 0 even though $c_{\mu\mu}^{(vw)} \neq 0$ (cf. Table 6.1).

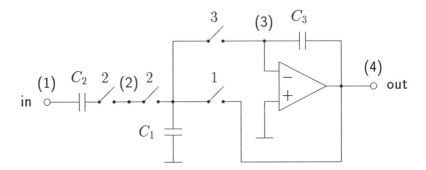

Fig. 6.23 Illustrative three-phase SC network.

between the nodes $V_\mu^{(w)}$ and $Q_{0\mu}^{(v)}$ that is generated by capacitor C_j is given by equation (6.20) (see also Table 6.1).

This brings us to the final observation that in the SFG of a general multiphase SC network, at most four branches, corresponding to a particular capacitor, can leave a source node $V_\mu^{(w)}$ (Fig. 6.22). This is because, in addition to the three possible branches leaving a source node $V_\mu^{(w)}$ in an SSN-type network, as shown Fig. 6.14, a fourth branch with the transfer function $c_{\mu\mu}^{(vw)}$ may exist from the source node $V_\mu^{(w)}$ to the sink node $Q_{0\mu}^{(v)}$ corresponding to the general node (g).

The method presented for the derivation of SFGs for general multiphase SC networks can be summarized as follows.

- For any capacitor charged in a particular switching phase from a particular source (or general node), all other general nodes are treated as sink nodes; the corresponding SFG branch transfer functions follow from the theory for SSN-type networks (section 6.2.3).

- Each capacitor charged in a particular switching phase μ from a particular general node (g) generates an SFG branch starting at the source node $V_\mu^{(w)}$ and ending in the sink node $Q_{0\mu}^{(v)}$ associated with that general node in μth phase. The transfer function of this branch is determined by equation (6.20) and given in Table 6.1.

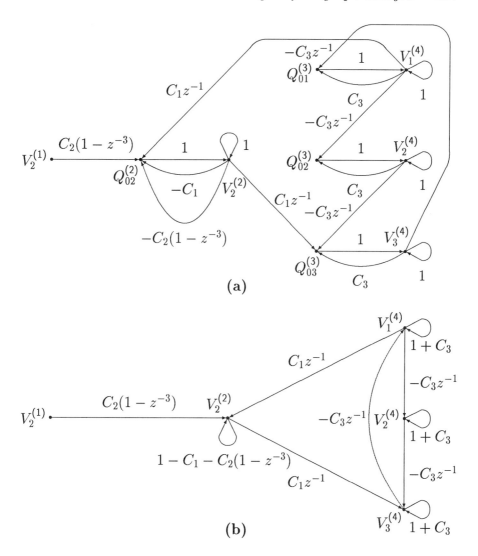

Fig. 6.24 SFGs corresponding to the illustrative three-phase SC network: (a) original SFG, (b) simplified SFG.

To illustrate the above by-inspection SFG derivation procedure, consider the three-phase SC network (Bedrosian and Refai, 1982) shown in Fig. 6.23. The switching phases are numbered from 1 to $M = 3$, as is customary in the literature, and not from 0 to $M - 1 = 2$, as was done (for convenience) in the theoretical presentation above. Note that the original single switch between capacitors C_1 and C_2 at node (2) has been split into two equivalent switches in series. This helps to visualize and localize the general node (2) (in practice this step may be left out). Besides the general node (2), the circuit contains two source nodes, (1) and (4), and one sink node, (3). The general node (2) is connected to the circuit only in phase 2 and need be represented in the SFG only in this phase, namely by the sink node $Q_{02}^{(2)}$ and the source node $V_2^{(2)}$. The corresponding SFG is shown in Fig. 6.24a. Note that, as a consequence of SSN-network transformation, nodes $V_2^{(2)}$ and $Q_{02}^{(2)}$ are connected by branches with transfer functions $-C_1$ and $-C_2(1-z^{-3})$ (cf. Table 6.1). The transfer function $-C_1$ follows from equations (6.16a) and (6.20), since the capacitor C_1 is discharged in phase 1. No such discharging phase exists for the capacitor C_2, and therefore the respective transfer function, $-C_2(1-z^{-3})$, is obtained from equations (6.16b) and (6.20) with $M = 3$. Fig. 6.24b shows the simplified SFG, i.e. that obtained after tying the sink nodes (representing the signals $Q_{02}^{(2)}$, $Q_{01}^{(3)}$, $Q_{02}^{(3)}$, $Q_{03}^{(3)}$ to the corresponding source nodes (i.e. nodes $V_2^{(2)}$, $V_1^{(3)}$, $V_2^{(3)}$ and $V_3^{(3)}$, respectively). Substituting $z^3 \to z$, this SFG agrees with that published by Bedrosian and Refai (1982). Using Mason's topological formula we obtain any desired transfer function, such as the voltage transfer function

$$H_{22}(z) = \frac{V_2^{(4)}}{V_2^{(1)}} =$$

$$\frac{C_1C_2(z^{-3}-1)z^{-3}}{C_2C_3z^{-6} + [C_1(C_1-C_3) - 2C_2C_3]z^{-3} + C_3(C_1+C_2)}.$$

7
FIR switched-capacitor filters

Finite impulse response (FIR) filters are the most popular structures for filtering of digital signals. It is well known that they can be designed for a perfectly linear phase characteristic and are inherently stable (e.g. Oppenheim and Schafer, 1975). They can, however, also be successfully realized using various analog techniques. Switched-capacitor realizations are quite important among them (Franca, 1985; Matsui et al., 1985; Fischer, 1987; Lee and Martin, 1988; Fischer, 1990; Korzec and Ciota, 1991; Dąbrowski, Menzi and Moschytz, 1992; Napieralski, Noullet and Ciota, 1994). FIR SC filters find applications in many aspects of signal processing. Sampling rate alteration is perhaps one of the most important of their applications (Martins and Franca, 1989, 1991; Ciota, Napieralski and Noullet, 1993) but they can also be used to realize various tasks in telecommunications, e.g. correlation computation or receiving of MFSK (minimum frequency-shift keying) signals (Dąbrowski, Menzi and Moschytz, 1992). Adaptive systems are also a very important application of FIR SC filters (Menzi, Zbinden and Moschytz, 1991; Menzi, 1992; Zbinden and Dąbrowski, 1992).

A digital filter realization comprises combinations of three types of elements, namely, delays, multipliers and summers. Since all these elements can be implemented in SC circuitry, any digital filter structure can, in principle, also be realized using SC techniques. Many such direct realizations would, however, be of little technical concern because of their costly implementation (large number of op amps, large number of capacitors, etc.). Multiphase realizations and op amp multiplexing are a good means of reducing the number of active elements (Fischer, 1990; Dąbrowski, Menzi and Moschytz, 1992). For some applications, e.g. adaptive filters, however, struc-

170 FIR switched-capacitor filters

tures with a reduced number of clock phases and a reduced number of capacitors, and thereby with an increased number of op amps, are more suitable (Zbinden and Dąbrowski, 1992).

The FIR SC filters presented in this chapter comprise the following building-blocks: sample-and-hold circuits, recharge memory elements, delay elements, summer circuits and rotator switches. The most popular basic circuit topologies for FIR SC filters are described. Furthermore, composite SC filter structures are derived using a morphological approach. The filter structures obtained are compared and evaluated with respect to the required chip area when implemented in 3 μm CMOS technology. Then we provide an overview of the most popular circuit architectures for multirate SC systems. First, the design of FIR decimators and interpolators with a single op amp is considered. Such structures are relatively simple but are subject to important limitations concerning maximum frequency range and acceptable integrated circuit complexity. Therefore, we also consider SC multirate circuits based on multi-amplifier architectures. Such structures are characterized by much more favourable requirements with respect to the number of capacitors and switches, and the complexity of the clock.

7.1 BUILDING-BLOCKS FOR FIR SC FILTERS

From the well-known expression for the FIR transfer function of order N

$$H(z) = \sum_{n=0}^{N} h_n z^{-n} , \qquad (7.1)$$

two basic FIR filter structures follow, as shown in Fig. 7.1: the **tapped delay line** structure and the **reversed delay line** structure. Both of them are based on a delay line concept. In the former (Fig. 7.1a), a conventional tapped delay line is used. Reversion of signal flow leads to the latter, which is depicted in Fig. 7.1b.

Modification of the operation of memory elements results in two additional, alternative FIR filter structures, as illustrated in Fig. 7.2. In the **parallel** structure in Fig. 7.2a, memory elements with increasing delays are connected in parallel. The **rotator** structure

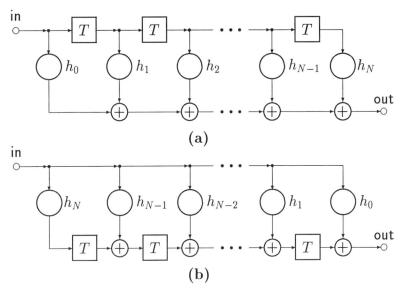

Fig. 7.1 Basic delay line FIR filter structures: (a) tapped delay line structure, (b) reversed delay line structure.

in Fig. 7.2b comprises a 'rotating' ring built with sample-and-hold elements.

Each one of the structures in Figs 7.1 and 7.2 can be used to implement FIR SC filters and in all cases only three elementary building-blocks are necessary to realize the filters, namely

- memory elements, which store a signal sample over one or more clock periods
- multipliers, which perform the multiplication of signal samples by a constant coefficient
- summers, which add two or more processed signal samples.

In what follows, we describe suitable SC circuit structures for these three operations.

7.1.1 Memory elements

There are essentially three ways of storing a signal sample in an SC circuit, corresponding to just discussed three concepts of con-

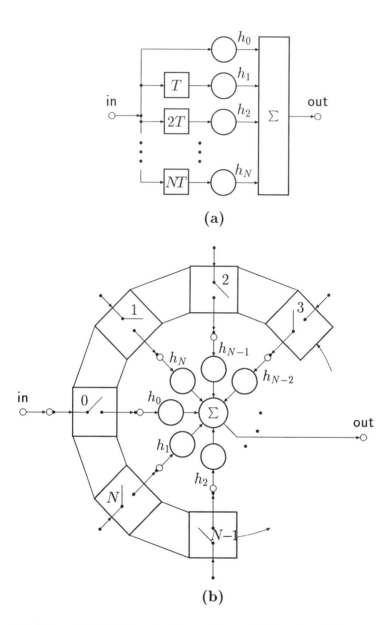

Fig. 7.2 Alternative FIR filter structures: (a) parallel structure, (b) rotator structure.

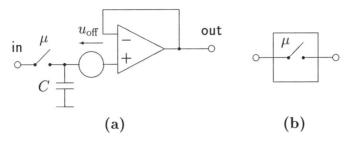

Fig. 7.3 SC sample-and-hold element: (a) SC circuit, (b) element symbol.

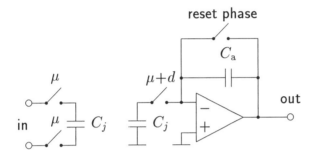

Fig. 7.4 Recharge memory element.

struction of FIR SC filters (rotator structure, parallel structure and delay line structure), resulting in the following building-blocks.

- **Sample-and-hold element** (used, for example, in a rotator structure), which stores a signal sample in the μth clock phase as a voltage on a capacitor (Fig. 7.3) and then transmits this voltage over a buffer so that it can be used in a later phase ν (before phase μ of the next operation period). The output voltage of the circuit in Fig. 7.3 is given by

$$u_{\text{out}}(\nu T) = \frac{1}{1+1/A}(u_{\text{in}}(\mu T) + u_{\text{off}}) \approx u_{\text{in}}(\mu T) + u_{\text{off}} \quad (7.2)$$

where A is the finite op amp DC gain. A symbol used for a recharge memory element is shown in Fig. 7.3b.

- **Recharge element** (applied in a parallel strucure and in delay line strucures), which stores a signal sample as a charge on a storage capacitor C_j in the μth clock phase (Fig. 7.4). For further processing, this charge is transmitted to the op amp feedback (or accumulator) capacitor C_a in phase $(\mu+d) \bmod M$. An important advantage of the recharge principle is that storage, multiplication and addition* of data are realized in the same building-block using only one op amp. However, the disadvantage lies in the fact that the stored sample can be processed only once, since it disappears after 'reading' or discharge of capacitor C_j. This sample is then accumulated in the feedback capacitor C_a as its charge increases and, therefore, it also manifests itself as the increase of the output voltage. This voltage can be reset to zero in some switching phase (when needed) with a reset switch.

- **Delay element** (a delay line building-block), which not only stores a new signal sample in the μth phase but also, in the same phase, transmits the old sample to the next delay element, thereby realizing a delay line. Such a delay element must contain at least one capacitor in addition to that of a sample-and-hold element.

(a) Recharge elements

The op amp output voltage of the recharge element in Fig. 7.4 is the product of the input voltage (delayed by d switching phases) and a coefficient equal to $h_j = \pm C_j/C_a$ (e.g., the jth coefficient of an FIR filter), where the sign of this coefficient depends on the switch configuration as discussed below. Thus, the transmission factor of such an element equals $h_j z^{-d}$ and can, therefore, be used directly for the realization of an FIR filter transfer function (7.1).

Consider the simple passive SC circuits shown in Fig. 7.5, connected between a source node on the left-hand side and a sink node on the right-hand side. We assume that this is the sink node that occurs in switching phase $\mu+d$ at the inverting input of the op amp in Fig. 7.4a. The circuits in Fig. 7.5 can be used as building-blocks

---

*In a single switching phase not only one but many storage capacitors $C_j, j = 1, 2, \ldots$, may be discharged on to the same feedback capacitor C_a.

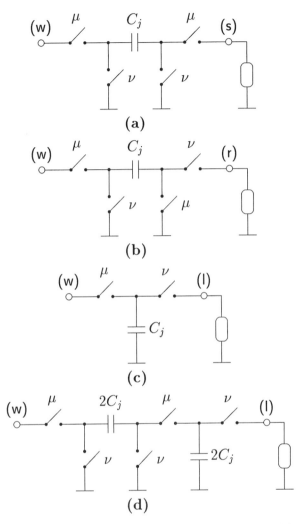

Fig. 7.5 Building-blocks for the recharge memory element: (a) open-floating-resistor (OFR) branch, (b) toggle-switch-inverter (TSI) branch, (c) toggle-switched-capacitor (TSC) branch, (d) parasitic-compensated-toggle-switched-capacitor (PCTSC) branch.

for the recharge memory element. A useful unified symbol for each of these circuits is depicted in Fig. 7.6.

The first circuit (Fig. 7.5a) is referred to as the **open-floating-resistor** (OFR) branch because it simulates a resistor T/C_j (cf.

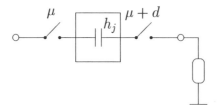

Fig. 7.6 Unified symbol for building-blocks of the recharge memory element.

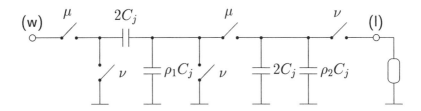

Fig. 7.7 PCTSC branch with indication of parasitic elements $\rho_1 C_j$ and $\rho_2 C_j$.

equation (1.3)). Using data listed in Table 6.1, we conclude that the recharge element with the OFR branch realizes a negative filter coefficient but with no delay, i.e. $d = 0$

$$h_j z^{-d} = -\frac{c_{j\mu\mu}^{(\text{sw})}}{C_\text{a}} = -\frac{C_j}{C_\text{a}} \ . \tag{7.3a}$$

The second element, i.e. that in Fig. 7.5b, is called the **toggle-switch-inverter** (TSI) branch. This branch realizes positive coefficients for arbitrary delays $d = \xi = (\nu - \mu) \bmod M$ (cf. expression (6.17) and Table 6.1)

$$h_j z^{-d} = -\frac{c_{j\nu\mu}^{(\text{rw})}}{C_\text{a}} = \frac{C_j}{C_\text{a}} z^{-\xi} \ . \tag{7.3b}$$

Negative coefficients for an arbitrary delay d can be realized by a conventional **toggle-switched-capacitor** (TSC) branch in Fig. 7.5c. Using Table 6.1 again, we get in this case

$$h_j z^{-d} = -\frac{c_{j\nu\mu}^{(\text{lw})}}{C_\text{a}} = -\frac{C_j}{C_\text{a}} z^{-\xi} \ . \tag{7.3c}$$

Unfortunately, the TSC branch is sensitive to the top-plate stray capacitance and can, therefore, lead to errors in the filter coefficients. However, the influence of the parasitic effects can be significantly reduced if the TSC branch is replaced by the more elaborate equivalent circuit, called the **parasitic-compensated-toggle-switched-capacitor** (PCTSC) branch (Franca, 1985), depicted in Fig. 7.5d. This circuit is redrawn in Fig. 7.7 in order to indicate parasitic elements $\rho_1 C_j$ and $\rho_2 C_j$. The total charge transferred per cycle (i.e. from phase μ to phase ν) by this circuit is

$$\begin{aligned} Q_{j\nu}^{(!)} &= \frac{2C_j + \rho_2 C_j}{4C_j + (\rho_1 + \rho_2)C_j} 2C_j z^{-\xi} V_\mu^{(w)} \\ &= \left(1 + \frac{\rho_2 - \rho_1}{4 + \rho_1 + \rho_2}\right) C_j z^{-\xi} V_\mu^{(w)} \\ &\approx c_{j\nu\mu}^{(lw)} V_\mu^{(w)} \ . \end{aligned} \quad (7.4)$$

With a proper layout, in which both capacitors $2C_j$ must exhibit identical geometries, the resulting coefficient error should become negligibly small.

Many building-blocks of the type in Fig. 7.5 may share one and the same op amp in Fig. 7.4. Thus, the just presented analysis of a recharge memory element is valid also for a recharge summer circuit, as discussed in section (b) below. The only difference is that, in the present case, the recharge takes place after d switching phases, rather than in the immediate next one, for the recharge summer circuit.

(b) Delay elements

Many different SC delay elements have already been proposed (Pain, 1979; Mulawka, 1981; Enomoto, Ishihara and Yasumoto, 1982; Grünigen, 1983; Gillingham, 1984; Nagaraj, 1984; Said, 1985; Dias and Franca, 1988; Dąbrowski, Menzi and Moschytz, 1989; Fischer, 1990). We shall here consider the even–odd delay element (Dąbrowski, Menzi and Moschytz, 1989) and the Gillingham delay element (Gillingham, 1984), both of which were found to be very suitable for many applications of FIR SC filters. The Nagaraj delay element (Nagaraj, 1984) has already been described and analysed in

178   *FIR switched-capacitor filters*

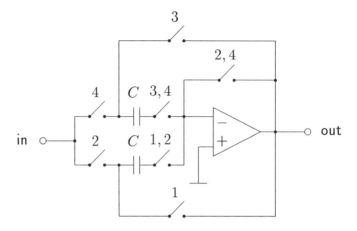

Fig. 7.8 Offset-compensated (odd) delay element.

section 6.1 (see Fig. 6.2). The question of suitability is related to the most important requirements imposed by the FIR tapped delay line, which are:

- minimum offset voltage, i.e. the possibility for offset compensation

- insensitivity to capacitor mismatch and to stray capacitances

- simple hardware, i.e. a small number of active elements and few clock phases per sample and/or delay line structure.

Even–odd delay element

An offset compensated, stray insensitive and capacitor mismatch insensitive delay element, the so called **even–odd delay element**, has been proposed (Dąbrowski, Menzi and Moschytz, 1989). Two versions of this circuit exist, namely the even delay element (Fig. 6.4) and the odd delay element (Fig. 7.8). Both versions require a four-phase clock but only two clock phases per sample. The input samples of the even element occur in clock phases 1 and 3, and the output samples in clock phases 2 and 4. By contrast, the input samples of the odd element occur in clock phases 2 and 4, and the

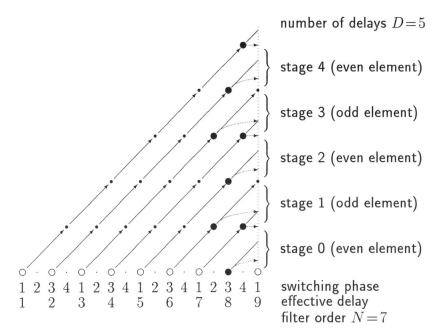

Fig. 7.9 Illustration of sample flow along a delay line with even and odd delay elements.

output samples in clock phases 1 and 3. Thus, in order to construct a delay line with these elements, both types must be alternately connected to each other. The flow of signal samples along such a delay line is illustrated in Fig. 7.9 for a line composed of five even–odd delay elements. The line contains nine effective delays of one sampling-period duration (one sampling period is equal to two clock phases). Generally, the number of even–odd elements D and the line length (i.e. the number of effective delays) N are related to each other by the following expression

$$D = \left\lfloor \frac{2N+1}{3} \right\rfloor \tag{7.5}$$

in which operation $\lfloor \cdot \rfloor$ denotes truncation. Thus, a delay line with these elements will actually contain less than the equivalent of 2/3 of an op amp per effective delay unit.

180   *FIR switched-capacitor filters*

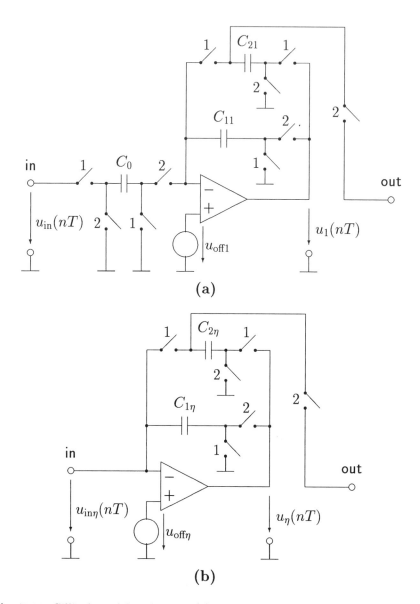

Fig. 7.10 Gillingham delay element: (a) circuit for the first stage of a delay line, (b) circuit for the following delay line stages.

Gillingham delay element

A two-phase stray insensitive SC delay circuit has been published by (Gillingham, 1984). There are two versions of this circuit, namely the circuit for the first stage of a delay line, comprising an input capacitor C_0 and two storage capacitors C_{11} and C_{21} (Fig. 7.10a) and the circuit for the following delay line stages, say for the stage η, containing only two storage capacitors $C_{1\eta}$ and $C_{2\eta}$ (Fig. 7.10b). The sampling period in this delay element is equal to two clock phases (as for an even–odd delay element). Assume that the discrete time instant nT (T being the duration of one clock phase) corresponds to switching phase 1. In the following analysis we assume ideal op amps (i.e. op amps with infinite gain) because effects related to finite op amp gain are much smaller than those related to other nonideal parameters. Thus, for the circuit in Fig. 7.10a, we obtain

$$u_1\big((n+1)T\big) = \frac{C_0}{C_{11}} u_{\text{in}}(nT) + u_{\text{off1}} \qquad (7.6a)$$

and

$$u_1\big((n+2)T\big) = \frac{C_{11}}{C_{21}} u_1\big((n+1)T\big) + u_{\text{off1}} - u_{\text{off2}} \, . \qquad (7.6b)$$

Similarly, for the circuit in Fig. 7.10b, taking into account that

$$u_{\text{in}\eta}(nT) = u_{\text{off}\eta} \, ,$$

we obtain

$$u_\eta\big((n+2\eta-1)T\big) = \frac{C_{2(\eta-1)}}{C_{1\eta}} u_{\eta-1}\big([n+2(\eta-1)]T\big) - u_{\text{off}(\eta-1)} + u_{\text{off}\eta} \qquad (7.7a)$$

and

$$u_\eta\big((n+2\eta)T\big) = \frac{C_{1\eta}}{C_{2\eta}} u_\eta\big((n+2\eta-1)T\big) + u_{\text{off}\eta} - u_{\text{off}(\eta+1)} \, . \qquad (7.7b)$$

Assuming that all capacitors in a delay line are equal, i.e. that

$$C_0 = C_{11} = C_{21} = \ldots = C_{1\eta} = C_{2\eta} = \ldots = C \, ,$$

we finally obtain

$$u_\eta\big((n+2\eta-1)T\big) = u_{\text{in}}(nT) + u_{\text{off1}} \qquad (7.8a)$$

and

$$u_\eta\big((n+2\eta)T\big) = u_{\text{in}}(nT) + u_{\text{off1}} + u_{\text{off}\eta} - u_{\text{off}(\eta+1)} \ . \qquad (7.8\text{b})$$

From the above equations, we conclude that the described Gillingham delay element is sensitive to capacitor mismatch and is not completely offset-voltage compensated. Nevertheless, the offset errors do not accumulate along the delay line realized as a cascade starting with the element in Fig. 7.10a and continuing with the elements in Fig. 7.10b.

Assume that we observe signal samples in switching phase 1. The Gillingham element in Fig. 7.10a performs a delay of the input signal by one sampling period, i.e. by two clock phases. This element can, however, be readily generalized to obtain a delay of, say, $(L-1)$ sampling periods, or more precisely $(L-1)$ delays of one sampling period, simply by using an L-phase rather than a two-phase clock and by extending the number of storage capacitors to L. By this means, an alternative realization for a delay line results. Such a line comprises $(L-1)$ delay stages but only a single active element, the multiplexed op amp driven by an L-phase clock (Fischer, 1990). Such a delay line is illustrated in Fig. 7.11.

7.1.2 Multiplier and summer circuits

Multiplication and addition can be obtained with a single SC building-block, often called a **summer circuit** (Allen and Sanchez-Sinencio, 1984). Exceptions are adaptive SC filters comprising varying, controlled multipliers that must, therefore, be separate blocks (Zbinden and Dąbrowski, 1992; Dąbrowski, 1993). We consider two types of SC summer circuit: the Lee–Martin summer circuit (Fig. 7.12a) (Lee and Martin, 1988) and a summer circuit based on the recharge principle (Fig. 7.12b) (Dąbrowski, Menzi and Moschytz, 1992), which we call a **recharge summer circuit**.

(a) Lee–Martin summer circuit

Lee and Martin (1988) proposed a two-phase stray insensitive SC summer circuit which is shown in Fig. 7.12a. To analyse this cir-

Building-blocks for FIR SC filters   183

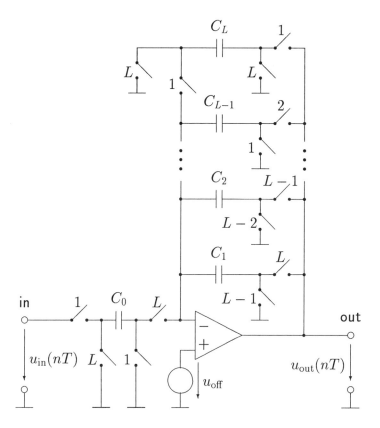

Fig. 7.11 A generalized Gillingham delay element realizing an L-phase/$(L-1)$-stage delay line.

cuit we assume that the discrete time instant nT corresponds to switching phase 1. Moreover, for convenience, we use the following terminology

$$C^- = \sum_{i=1}^{K} C_i \quad \text{and} \quad C^+ = \sum_{i=K+1}^{L} C_i \qquad (7.9)$$

and

$$u_\infty(nT) = -\sum_{i=1}^{K} \frac{C_i}{C_\text{sum}} u_i(nT) + \frac{C_\text{sum} + C^-}{C_\text{add} + C^+} \sum_{i=K+1}^{L} \frac{C_i}{C_\text{sum}} u_i(nT) \;. \qquad (7.10)$$

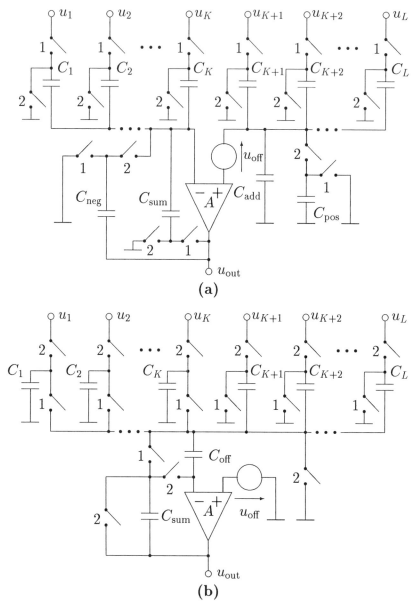

Fig. 7.12 SC summer circuits: (a) Lee–Martin summer, (b) recharge summer.

Simple analysis yields the following expression for the output voltage

$$u_{\text{out}}(nT) = \frac{C_{\text{sum}}}{C_{\text{sum}} + \dfrac{1}{A}(C_{\text{sum}} + C^-)} u_\infty(nT)$$

$$+ \frac{C_{\text{sum}} + C^-}{(A+1)C_{\text{sum}} + C^-} u_{\text{out}}\big((n-1)T\big) \quad (7.11a)$$

and

$$u_{\text{out}}\big((n+1)T\big) =$$

$$\frac{\left[C_{\text{neg}} + C_{\text{sum}} + \dfrac{1}{A}(C_{\text{sum}} + C^-)\right] u_{\text{out}}(nT) - C_{\text{sum}} u_\infty(nT) + C_{\text{neg}} u_{\text{off}}}{C_{\text{neg}} + \dfrac{1}{A}(C_{\text{neg}} + C_{\text{sum}} + C^-)}.$$

$$(7.11b)$$

It is convenient to choose the capacitance so that

$$C_{\text{sum}} + C^- = C_{\text{add}} + C^+ . \quad (7.12)$$

With the additional assumption that $A \to \infty$, equations (7.10)–(7.11b) reduce to

$$u_{\text{out}}(nT) = u_\infty(nT) = -\sum_{i=1}^{K} \frac{C_i}{C_{\text{sum}}} u_i(nT) + \sum_{i=K+1}^{L} \frac{C_i}{C_{\text{sum}}} u_i(nT)$$
$$(7.13a)$$

and

$$u_{\text{out}}\big((n-1)T\big) = u_{\text{out}}(nT) + u_{\text{off}} . \quad (7.13b)$$

From equations (7.13a) and (7.13b), we conclude that the Lee–Martin summer circuit is offset-compensated, but only in the discrete-time instants corresponding to switching phase 1.

(b) Recharge summer circuit

Assuming again that the discrete time instant nT corresponds to switching phase 1, we obtain, for the recharge summer circuit shown in Fig. 7.12b

186 FIR switched-capacitor filters

$$u_{\text{out}}(nT) = \frac{-\sum_{i=1}^{K} C_i u_i((n-1)T) + \sum_{i=K+1}^{L} C_i u_i((n-1)T) + \frac{1}{1+A}\left(C_{\text{sum}} + \sum_{i=1}^{L} C_i\right) u_{\text{off}}}{C_{\text{sum}} + \frac{1}{A}\left(C_{\text{sum}} + \sum_{i=1}^{L} C_i\right)}.$$
(7.14)

Neglecting the effects of finite op amp gain A as before, we obtain

$$u_{\text{out}}(nT) = -\sum_{i=1}^{K} \frac{C_i}{C_{\text{sum}}} u_i((n-1)T) + \sum_{i=K+1}^{L} \frac{C_i}{C_{\text{sum}}} u_i((n-1)T) . \quad (7.15)$$

Note that this structure is not completely top-plate stray insensitive, since a toggle-switch capacitor (TSC) configuration is used for summation with negative coefficients. The influence of the resulting parasitic capacitance may, however, be compensated for by replacing the TSC branches by slightly more complicated PCTSC branches (Fig. 7.5d).

7.2 MORPHOLOGICAL DESIGN OF FIR SC FILTERS

In this section we introduce basic FIR SC filter structures using the elementary building-blocks presented in section 7.1. Furthermore, using the so called 'morphological approach' (Zwicky, 1967; Moschytz, 1976), we derive composite FIR SC filter structures composed of different basic structures. The resulting FIR SC filter structures are then compared and evaluated.

7.2.1 Basic FIR SC filter structures

Using the three types of SC memory element, given in section 7.1.1, we obtain three different basic FIR SC filter structures.

- **Delay line FIR structure** (Fig. 7.1), in which signal samples are successively transmitted from one storage or delay element to the next and all elements over the whole delay line must be modified in each operation cycle

- **Parallel** or in other words **multi-C FIR structure** (Fig. 7.2a), which is a combination of recharge memory elements

- **Rotator FIR structure** (Fig. 7.2b), in which memory is obtained by sample-and-hold elements.

Delay line FIR configurations follow directly from the expression for an FIR filter transfer function (7.1). The resulting filter topologies can be based either on the tapped delay line structure (Fig. 7.1a) or on the reversed delay line structure (Fig. 7.1b). Both of them can, in turn, be realized in parallel or serially. In the first case, 'data read' (and then 'data write') operations are performed in parallel, i.e. in the same phase in all delay elements. As a result, a small number of clock phases but large number of active elements are necessary (Dąbrowski, Menzi and Moschytz, 1992). Such a realization is, therefore, referred to as the **multi-op-amp delay line structure**. In the second case, all operations are realized serially in a single filter operation cycle. Therefore, only one, sufficiently quick (multiplexed) op amp occurs in such a realization but a complex multiphase clock is necessary (Fischer, 1985, 1990; Ciota, 1996). This realization is referred to as the **multiphase delay line structure**.

Various compromise realizations between these two extremes are possible, comprising a moderate number of op amps and clock phases (see, for instance, a modified four-phase Gillingham delay element in Fig. 7.19).

Multi-op-amp delay line structures may be formed either with even-odd delay elements or with Gillingham delay elements connected to a summer circuit. In the first case, however, the recharge summer circuit must be used, because the signal samples to be summed occur in different switching phases (cf. Fig. 7.9). This realization, called the delay line FIR structure No. 1, is shown in Fig. 7.13. With the Gillingham delay elements, both the Lee–Martin summer and the recharge summer circuits can be used. Thus, we obtain two additional realizations which we call the delay line FIR structure No. 2 and the delay line FIR structure No. 3, respectively (Fig. 7.14). The number of circuit elements necessary to build each of these three structures was computed by Dąbrowski, Menzi and Moschytz (1992). All of them are characterized by a very low number of switching phases, but require relatively many

188 *FIR switched-capacitor filters*

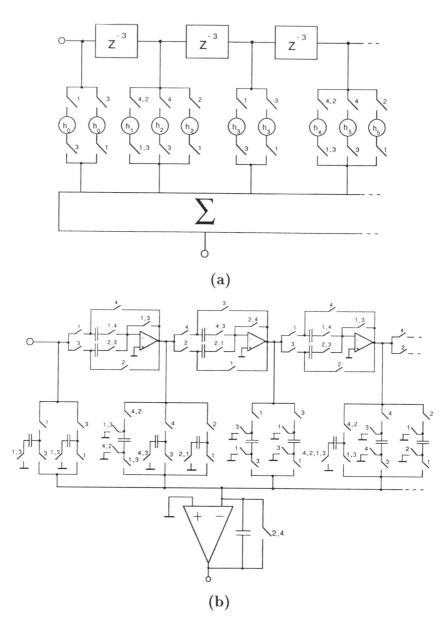

Fig. 7.13 Delay line structure No. 1: (a) simplified illustrative scheme of order 5, (b) corresponding circuit realization; coefficients h_0, h_2 and h_4 are assumed to be negative, coefficients h_1, h_3 and h_5 are assumed to be positive.

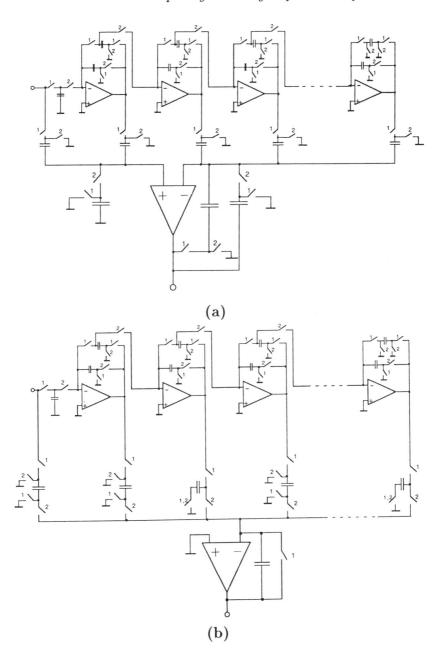

Fig. 7.14 Delay lines with Gillingham delay elements: (a) delay line structure No. 2, (b) delay line structure No. 3.

190 FIR switched-capacitor filters

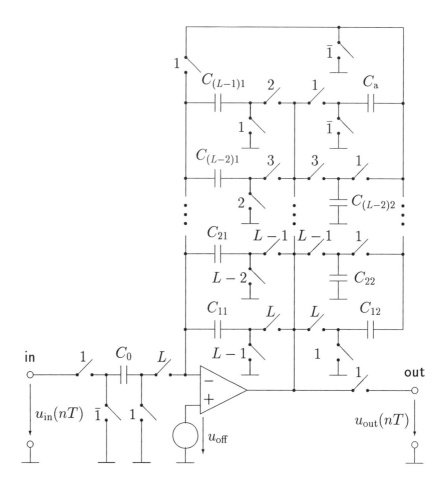

Fig. 7.15 Realization of an FIR filter of order $(L-2)$ using L-phase delay line structure No. 4.

op amps. Therefore, multiphase delay line structures have been proposed (Fischer, 1990). In Fig. 7.15 an L-phase$/(L-2)$th-order structure with a tapped delay line is depicted. This type of configuration is called the delay line FIR structure No. 4. In Fig. 7.15 we assumed that the number L is even and that coefficients $h_0, h_2, \ldots, h_{L-2}$ are positive, while coefficients $h_1, h_3, \ldots, h_{L-3}$ are negative. The corresponding transfer function is given by

Morphological design of FIR SC filters

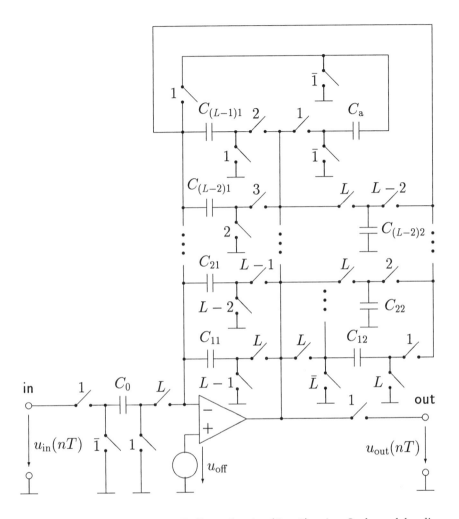

Fig. 7.16 Realization of an FIR filter of order $(L-2)$ using L-phase delay line structure No. 5.

$$H(z) = z^{-1}(h_0 + h_1 z^{-1} + \ldots + h_{L-3} z^{-(L-3)} + h_{L-2} z^{-(L-2)}) =$$

$$\frac{C_0}{C_a} z^{-1} \left(\frac{C_{12}}{C_{11}} - \frac{C_{22}}{C_{21}} z^{-1} + \ldots - \frac{C_{(L-2)2}}{C_{(L-2)1}} z^{-(L-3)} + z^{-(L-2)} \right).$$

Slight modification of switching phases leads to the reversed delay line realization, called the delay line FIR structure No. 5 (Fig.

7.16). In this case, the corresponding transfer function is given by

$$H(z) = z^{-1}(h_0 + h_1 z^{-1} + \ldots + h_{L-3} z^{-(L-3)} + h_{L-2} z^{-(L-2)}) =$$

$$\frac{C_0}{C_a C_{11}} z^{-1} \left(C_{12} - C_{22} z^{-1} + \ldots - C_{(L-2)2} z^{-(L-3)} + C_{11} z^{-(L-2)} \right).$$

The main drawback of all delay line structures is their sensitivity to overwrite errors accumulating along the delay line as a signal is transmitted along the line. To reduce these errors, parallel and rotator structures may be used.

With the parallel FIR structure (called also the multi-C FIR structure), each signal sample is stored on individual capacitors corresponding to $N+1$ filter coefficients. In each phase of the filter, the respective capacitors are recharged over a feedback capacitor of the recharge summer circuit, by which means the filtering algorithm is performed. This structure makes it possible to reduce the number of op amps to only one, but the price for such a solution is a large number of capacitors and switching phases (Fig. 7.17).

A rotator FIR structure comprises a so-called rotator switch that connects signal samples (stored by sample-and-hold elements) with the capacitors of a summer circuit, thereby realizing the coefficients of the filter. A rotator FIR structure with a Lee–Martin summer circuit is shown in Fig. 7.18. An advantage of the rotator FIR structure is that no cumulative errors occur; however, the required number of op amps and switching phases is relatively large.

7.2.2 Composite FIR SC filter structures

Since each of the basic FIR SC filter structures discussed above possesses certain drawbacks which get worse as the filter length is increased, we can consider combinations of FIR SC structures composed of two or more basic structures of smaller length. Such combinations are, in fact, equivalent to the polyphase decomposition of the transfer function to be realized (cf. section 4.4.4). Thus, the obtained solutions correspond to the polyphase configurations shown in Fig. 4.14. Each polyphase component can then be further decomposed and so on, resulting in a multistage decomposition.

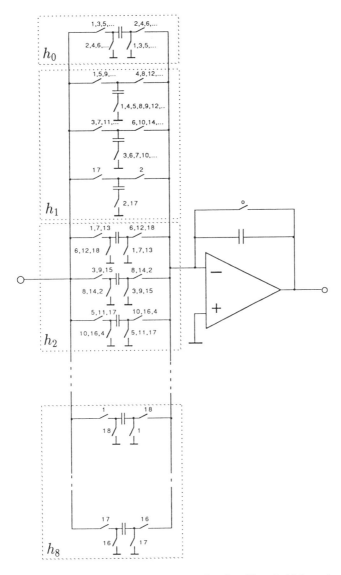

Fig. 7.17 Parallel (multi-C) FIR structure of order $N = 8$ (driven by 18-phase clock, $18 = 2(N + 1)$); coefficients $h_0, h_2, \ldots, h_8$ are assumed to be positive, coefficients $h_1, h_3, \ldots, h_7$ are assumed to be negative.

The single-stage decomposition is, however, of the greatest practical concern.

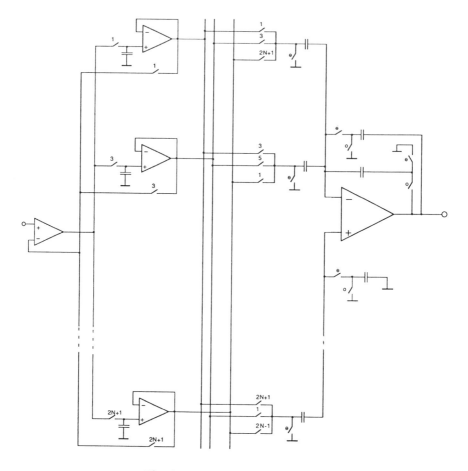

Fig. 7.18 Rotator FIR structure.

The particular importance of polyphase decomposition is obvious for multirate systems, since the polyphase components of such systems operate with a reduced sampling rate (cf. Figs 4.15 and 4.16) and can, therefore, be realized very efficiently.

To derive and evaluate all possible and technically reasonable compositions of the basic FIR SC filter structures, we follow the morphological approach (Zwicky, 1967; Moschytz, 1976). This approach consists of the following five steps.

Step 1 *Formulation of problem:* Realize composite FIR SC filter configurations using the basic FIR SC filter structures introduced in section 7.2.1. Find the best solutions with respect to the number of circuit elements, chip area, operation speed, network performance, clock complexity, etc.

Step 2 *Characterization of fundamental elements:* The following four basic FIR SC filter structures serve as the fundamental elements for the so-called 'morphological box':

- multi-op-amp delay line FIR structure

- multiphase delay line FIR structure

- parallel (multi-C) FIR structure

- rotator FIR structure.

Step 3 *Derivation of multidimensional matrix or morphological box:* In our case, the morphological box consists of a four-dimensional matrix, since we have four fundamental elements (four basic FIR SC filter structures). Each solution corresponds to a different combination of these elements, theoretically resulting in a total of 16 different solutions for one-stage decomposition, 64 solutions for two-stage decomposition, 256 solutions for three-stage decomposition and so on.

Step 4 *Evaluation of each solution contained in a morphological box:* From a practical realization point of view, many of the theoretically possible solutions contained in the morphological box can be eliminated immediately. The considerations involved in this elimination are as follows:

- In order to reduce the number of op amps, the multi-C FIR structure should be applied as the last link of a composite filter structure or followed by a delay line composed of recharge branches and recharge summer circuits. Signal samples stored in passive recharge branches can be read only once, so they may not serve as inputs to other links.

- A particular basic structure can, in practice, occur only once in a composite structure. Configurations composed of more than two basic structures are of little practical concern.

196   *FIR switched-capacitor filters*

Table 7.1 Chip area estimates for 3 μm CMOS technology

Element	Area ($10^3 \mu m^2$)
op amp	37.0
switch	1.0
unity capacitor	0.9
coefficient capacitor	~ 144.0

- Structures equivalent to a tapped delay line are suitable for the first link in the type 1 decomposition (Fig. 4.14a), while structures realizing the reversed delay line are well fitted for the last link of the type 2 decomposition in Fig. 4.14b.

Taking the above considerations into account, the great amount of theoretical combinations can be reduced to only several composite FIR SC filter structures of practical importance (Dąbrowski, Menzi and Moschytz, 1992).

Step 5 *Selection of best solutions:* The selection of the best of the FIR SC structures remaining in step 4 is based on (i) an evaluation of their feasibility as integrated circuits, and (ii) their performance in the particular application for which they were designed.

7.2.3   Evaluation of FIR SC filter structures

In order to evaluate the basic as well as the composite FIR SC filter structures, 12 filter structures have been chosen, on account of their relatively small chip area requirements. They were tested for the chip area necessary for their integration in 3 μm CMOS technology. The following FIR SC filter structures were considered:

1. delay line FIR structure No. 1
2. delay line FIR structure No. 2
3. delay line FIR structure No. 3
4. delay line FIR structure No. 4
5. delay line FIR structure No. 5

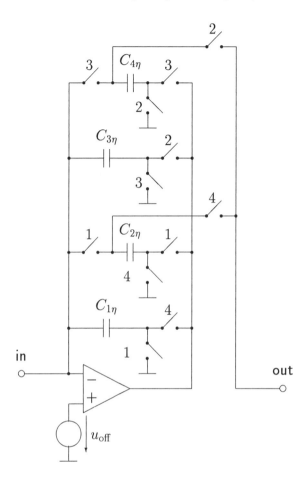

Fig. 7.19 Modified four-phase Gillingham delay element.

6. multi-C FIR structure

7. rotator FIR structure

8. rotator–multi-C FIR structure

9. rotator–delay line No. 1, 2 or 3 FIR structure

10. multi-C–delay line FIR structure, or, in other words, active-delayed block (ADB) structure (Franca and Santos, 1988)

198   *FIR switched-capacitor filters*

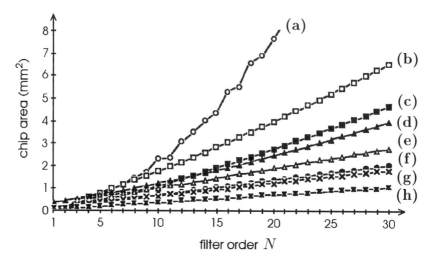

Fig. 7.20 Chip area for different FIR filter structures as a function of filter order N: (a) multi-C, (b) rotator–delay line, (c) rotator–multi-C, (d) rotator, (e) delay line–rotator, (f) delay lines No. 1, 2 and 3, (g) delay line–multi-C and multi-C–delay line (ADB), (h) delay lines No. 4 and 5.

11. delay line No. 1, 2 or 3–rotator FIR structure

12. delay line No. 1, 2 or 3–multi-C FIR structure.

For the past two types of composite structures, modified four-phase Gillingham delay elements were used (Fig. 7.19). Furthermore, the Lee–Martin summer circuit was used in the delay line–rotator FIR structure and the recharge summer circuit in the delay line–multi-C FIR structure and in the active-delayed block (ADB) structure.

The resulting chip area, as a function of filter order N, is plotted for above test FIR structures in Fig. 7.20. This evaluation is based on the chip area estimates listed in Table 7.1. From Fig. 7.20 we conclude that the most efficient structures, at least from the chip area point of view, are delay line structures No. 4 and No. 5 and also composite structures: delay line–multi-C and multi-C–delay line (ADB structure). The less efficient is the multi-C structure driven by a standard multiphase clock. Its efficiency can, however, be improved as discussed in section 7.3.1.

7.3 MULTIRATE FIR SC FILTERS

The most efficient architectures for implementing FIR SC multirate systems are the polyphase structures introduced in section 4.4.4. The polyphase configuration depicted in Fig. 4.15 can be used to realize a decimator. On the other hand, the configuration shown in Fig. 4.16 is suitable for the polyphase realization of an interpolator. The simplest way to realize both these configurations with SC circuits is to adapt the multi-C structure that comprises two types of components, namely the transmission factors (i.e. the passive recharge branches in Fig. 7.5) and the accumulator (the op amp feedback capacitor C_a in Fig. 7.4).

7.3.1 FIR SC decimators

Assume that the multi-C structure in Fig. 7.17 realizes a polyphase component of the decimator in Fig. 4.15. Denote by N_ν the order of this component. Then such multi-C structure should be driven by a $2(N_\nu + 1)$-phase clock. To implement the first coefficient h_0, one passive recharge branch is necessary (the TSI branch for the positive coefficient and the TSC or PCTSC branch for the negative coefficient, cf. Fig. 7.5).

One could think that for the realization of the second coefficient h_1 we should need two recharge branches, for the third coefficient h_2 three recharge branches and so on, but this is only true if number $N_\nu + 1$ is even, divisible by 3 and so on, respectively. The example depicted in Fig. 7.17 is of order $N_\nu = 8$, thus, the additional, third, recharge branch (that operating between switching phases 17 and 2) is necessary for the realization of coefficient h_1. The third coefficient h_2 is also realized by three recharge branches because number $N_\nu + 1 = 9$ is divisible by 3 and $(N_\nu + 1)/3 = 3$.

Generally, the number $\varrho_{\nu\eta}$ of recharge branches realizing coefficient h_η, $\eta = 0, 1, \ldots, N_\nu$ is equal to

$$\varrho_{\nu\eta} = \eta + 1 + (N_\nu + 1) \bmod (\eta + 1) \qquad (7.16)$$

Thus the number ϱ_ν of all recharge branches in the νth polyphase component is given by

$$\varrho_\nu = \sum_{\eta=0}^{N_\nu} \varrho_{\nu\eta} = \frac{(N_\nu + 1)(N_\nu + 2)}{2} + \sum_{\eta=0}^{N_\nu} (N_\nu + 1) \bmod (\eta + 1) \quad (7.17)$$

and the number ϱ of recharge branches in the whole M-fold decimator can be computed as

$$\varrho = \sum_{\nu=0}^{M-1} \varrho_\nu \ . \quad (7.18)$$

The big number ϱ obtained is the main drawback of the multi-C structure, resulting in a large chip area (cf. Fig. 7.20a). It is, however, possible to reduce numbers $\varrho_{\nu\eta}$ to

$$\varrho_{\nu\eta} = \eta + 1 \ , \quad (7.19)$$

thus to reduce correspondingly the required chip area, by modification of clock pulses. Instead of the standard $2(N_\nu + 1)$-phase clock used for all recharge branches of all coefficients h_η, individual recharge branches should be driven by two-phase (i.e. second order) heteromerous clocks with successively decreasing average rates according to the increasing value of η.

Further reduction of chip area is possible with composite structures. An interesting solution with a multi-C–delay line structure, or, in other words, the active-delayed block (ADB) polyphase structure was proposed by Franca and Santos (1988).

7.3.2 FIR SC interpolators

The convenient architecture for implementing an M-fold SC interpolator is the polyphase configuration in Fig. 4.16. In the digital case, the sampling rate increase is achieved just by interleaving the samples produced by individual polyphase branches, as shown on the right-hand side of Fig. 4.16. In an SC circuit, however, we usually work with sampled-and-held signals, and therefore the simple interleaving operation must be replaced by distributed summing, as illustrated on the left-hand side of Fig. 4.16. It is certainly possible to implement such summing with only one shared accumulator (an op amp with the accumulating feedback capacitor C_a, operating in appropriate switching phases – cf. Fig. 7.4).

An important consequence of summing of sampled-and-held signals instead of sample interleaving is the modification of the resulting interpolator transfer function $H(z)$. Denote by $H_d(z)$ the original interpolator transfer function (that corresponding to the discrete-time circuit operation). Then the resultant transfer function is given by

$$H(z) = H_d(z) H_{\text{S\&H}}(z) \qquad (7.20)$$

where

$$H_{\text{S\&H}}(z) = \sum_{m=0}^{M-1} z^{-m} . \qquad (7.21)$$

An important particular class of interpolating circuits is constituted by the so-called **linear interpolators** (Neuvo, Dong and Mitra, 1984; Saramäki, Neuvo and Mitra, 1988). An M-fold linear interpolator generates $M - 1$ equally spaced and linearly interpolated samples between each pair of neighbouring original samples. Assume that $x_o(\tau_\nu)$ is the interpolator input signal defined by expression (4.1b), where $\tau_\nu = \tau_0 + \nu T_o$ and the input signal sampling period $T_o = MT$, T being the output signal sampling period. Then the output signal of the causal interpolator can be defined as[*]

$$x(t_n) = x(\tau_0 + nT) = x_o(\tau_{\nu-1}) + \frac{\mu}{M}\left[x_o(\tau_\nu) - x_o(\tau_{\nu-1})\right] \qquad (7.22)$$

where $n = \nu M + \mu$ and $\mu = 0, 1, \ldots, M - 1$.

We can now compute the corresponding transfer function $H(e^{j\omega T})$, treating signal $x_u(\tau_0 + nT)$, defined according to equation (4.3), as the hypothetical input signal to the system described by transfer function $H(e^{j\omega T})$ (cf. Fig. 4.1). Then

$$\begin{aligned} H(z) &= z^{-M} + \sum_{\mu=1}^{M-1} \left[z^{-M-\mu} + \frac{\mu}{M}\left(z^{-\mu} - z^{-M-\mu}\right) \right] \\ &= z^{-M} + \frac{1}{M} \sum_{\mu=1}^{M-1} \mu \left(z^{-\mu} + z^{-2M+\mu} \right) \end{aligned} \qquad (7.23)$$

---

[*]A causal interpolator could, in fact, induce an effective signal delay, that is a delay one period T less than that assumed in equation (7.22). This is, however, irrelevant in our analysis.

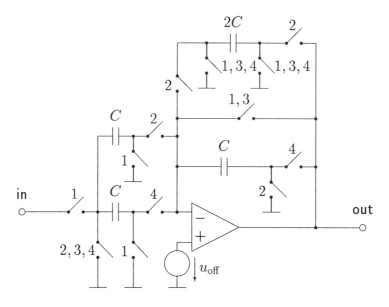

Fig. 7.21 A two-fold/four-phase linear interpolator.

where variable z is defined with respect to the sampling period T.

Linear interpolators can be realized using elementary SC techniques. As an example, a two-fold/four-phase linear interpolator putting an average value sample in between every two existing samples is shown in Fig. 7.21 (Ciota, 1996). The clock pulse-width T_c is related to periods T_o and T as follows

$$T_o = 2T = 4T_c \ .$$

The input signal is sampled in switching phase 1, while output samples occur in switching phases 2 and 4. The overall signal delay induced by this interpolator is from switching phase 1 to phase 4, i.e. by one clock pulse-width T_c less than the delay assumed in equations (7.22) and (7.23). Thus, the resultant transfer function $H(z)$ must be multiplied by a factor of $z^{0.5}$ with respect to that defined by expression (7.23), so is equal to

$$H(z) = 0.5z^{-0.5} + z^{-1.5} + 0.5z^{-2.5}$$

where variable z is defined with respect to the sampling period T.

8

IIR switched-capacitor filters

For certain applications, in particular for low-power and highly selective systems, **infinite impulse response** (IIR) circuits have distinct advantages over their FIR (finite impulse response) counterparts discussed in Chapter 7. This is because the filter order necessary for IIR realizations is in most cases substantially reduced compared to that corresponding to FIR filters.

Multirate IIR systems can be realized, in a similar way to FIR solutions, either by using direct structures, depicted for an interpolator and a decimator in Figs 4.1 and 4.3, respectively (such filters operating with a higher sampling rate), or by implementing the polyphase arrangements shown in Figs 4.15 and 4.16 (the polyphase component filters operating with the lower sampling rate). Mixed solutions in which a part of the filter is realized directly and the rest is decomposed into polyphase components are also possible and reasonable (Franca and Mitra, 1994). Recently, cascaded decimation structures have been proposed (Martins and Franca, 1995).

From the discussion in section 4.4.4 and in Chapter 7, we can conclude that in an FIR case, it is always advantageous to resolve upon the polyphase realization, owing to the existence of a very simple and efficient polyphase decomposition (4.56) for any FIR filter transfer function. This is, however, not the case for multirate IIR systems. Although the IIR polyphase decomposition is always feasible (cf. equations (4.60) and (4.61)) and in particular cases can be very efficient (e.g. if polyphase components are all-pass functions as in section 4.6.1), it generally leads to an increase in the overall system complexity. Thus, selection of the optimum multirate IIR arrangement for a particular application is not an easy question. For low and medium frequency range applications, i.e. if there is no problem with the op amp settling time, direct realizations and particularly those based on multirate ladder arrangements (section 4.6.2) with the recovery of the effective pseudoenergy (section 8.3) may be the optimum solution (Dąbrowski, 1987a).

Independently from the choice of the best multirate IIR arrangement for a given application, many different methods for the design of the corresponding IIR filter structures are possible. Among them two main classes may be distinguished:

- **direct methods** based on the realization of the discrete-time (i.e. $\mathcal{Z}$ domain) transfer function (e.g. Kaelin, 1988)

- **indirect methods** based on the reference circuit concept (section 3.2.2); the reference circuit is first designed, then the resulting SC circuit is obtained by the bilinear transformation (3.54).

Indirect methods can be divided into two separate subclasses depending on the type of correspondence between the reference circuit and the resulting SC circuit. Thus, two separate classes of SC networks exist:

- SC networks modelling voltage–charge relations in the reference circuit (Temes, Orchard and Jahanbegloo, 1978; Nossek and Temes, 1980; Lee et al., 1981; Dąbrowski, 1982b, 1985; Hökenek and Moschytz, 1983; Montecchi and Maloberti, 1983; Handkiewicz, 1988)

- SC networks modelling voltage waves in the reference circuit – these are the so-called **voltage inverter switch** SC **circuits** (designated as VIS-SC circuits for short) (Fettweis, 1979a, b) and wave-SC circuits (Kleine et al., 1981; Mavor et al., 1981). Wave-SC circuits are more complicated than VIS-SC circuits and will not be further considered.

The main advantage of the indirect methods based on the reference circuit concept lies in the fact that the resulting SC realizations preserve (if properly designed) the marvellous robust stability and small sensitivity features of their reference circuits.

8.1 DOUBLE-FREQUENCY TRANSFORMATION

We shall now describe the method for the design of IIR SC networks, based on the bilinear transformation and modelling the voltage–charge relations in the reference circuit (Dąbrowski, 1982b; Dąbrowski, 1985).

Let $H(\psi)$ and $H_{\mathrm{SC}}(z)$ be the synthesized reference circuit transfer function and the transfer function of the final SC realization, respectively. These functions are to be related bilinearly*, i.e.

$$H_{\mathrm{SC}}(z) = H\left(F_1(z)\right) \qquad (8.1)$$

where

$$\psi = F_1(z) = \frac{2}{T}\frac{z-1}{z+1} \qquad (8.2)$$

and T is the sampling period equal, e.g. to two clock pulses T_c in a two-phase SC circuit.

Notice first that any realization of the reference circuit can be transformed into an active-RC SSN-type circuit by the method similar to that described in section 6.2.4. Consider any resistor R_j in such a circuit. Denoting voltage across this resistor by $V_j(\psi)$ and the total charge by $Q_j(\psi)$, we obtain

$$V_j(\psi) = R_j \psi Q_j(\psi) \ . \qquad (8.3)$$

If we could now find an SC configuration realizing equation (8.3) with variable ψ substituted by bilinear function $F_1(z)$, the synthesis task would be solved simply by replacing each resistor by such a configuration. A simple SC configuration of this type was proposed by Temes, Orchard and Jahanbegloo (1978) but its application leads to stray-sensitive realizations. We therefore prefer to use SC building-blocks as depicted in Fig. 7.5 rather than that configuration. However, these blocks, instead of function $F_1(z)$, realize other functions denoted by $F_2(z)$ and listed in Table 8.1. Thus we

---

*Notice that we consider here the bilinear transformation $F_1(z)$ defined, for convenience, by equation (8.2). The difference from the bilinear transformation previously introduced in equation (3.54) is the factor $2/T$ corresponding merely to some frequency normalization. This difference is certainly of secondary importance.

206 IIR switched-capacitor filters

Table 8.1 Exemplary frequency transformations

Transformation type and SC building block in Fig. 7.5	$s = F_2(z)$	$z = F_2^{-1}(s)$	$\psi = F_3(s)$	s-plane image of the left ψ-halfplane
Forward Euler transformation TSC and PCTSC	$\dfrac{1}{T}(z-1)$	$1 + sT$	$\dfrac{2}{T}\dfrac{sT}{2+sT}$	circle in left half-plane, tangent to imaginary axis at origin, centered at $-\frac{2}{T}$
Reversed forward Euler transformation TSI	$\dfrac{1}{T}(1-z)$	$1 - sT$	$\dfrac{2}{T}\dfrac{sT}{sT-2}$	circle in right half-plane, tangent to imaginary axis at origin, centered at $\frac{2}{T}$
Backward Euler transformation OFR	$\dfrac{1}{T}\dfrac{z-1}{z}$	$\dfrac{1}{1-sT}$	$\dfrac{2}{T}\dfrac{sT}{2-sT}$	left half-plane minus circle in right half-plane tangent to imaginary axis at origin, centered at $\frac{2}{T}$

T – sampling period, i.e. two clock pulses T_c in a two-phase SC circuit.

can generally write

$$V_j(z) = R_j F_2(z) Q_j(z) \qquad (8.4)$$

for a resistor R_j replaced by any SC building-block of Fig. 7.5, driven by a two-phase clock.

The proposed SC circuit synthesis procedure consists in a double-frequency transformation and is composed of the following steps:

- choose an SC building-block described by equation (8.4) where $F_2(z)$ should be a one-to-one function

- transform function $H_{\text{SC}}(z)$ to the new s-domain

$$H_{\text{RC}}(s) = H_{\text{SC}}\left(F_2^{-1}(s)\right) = H\left(F_3(s)\right) \qquad (8.5)$$

where
$$F_3(s) = F_1\left(F_2^{-1}(s)\right) \tag{8.6}$$

- realize the function $H_{\mathrm{RC}}(s)$ by an active-RC SSN-type circuit – this circuit is called the **secondary reference circuit** or the **secondary prototype** in contradistinction to the 'primary' reference circuit realizing the transfer function $H(\psi)$

- replace all resistors R_j, $j = 1, 2, \ldots$, of the secondary prototype by the chosen SC building-blocks, which result in the final SC realization.

From relation (8.6) it follows that the above synthesis procedure is based on the composition of two transformations, namely $F_2(z)$ and $F_3(s)$ resulting in the bilinear mapping (8.2)

$$F_1(z) = F_3\big(F_2(z)\big) \ . \tag{8.7}$$

Exemplary functions $F_2(z)$ and $F_3(s)$ are presented in Table 8.1. They are termed according to the adequate numerical method for solving initial-value problems for ordinary differential equations. The transformation in the second row of Table 8.1 is called the 'reversed forward Euler transformation', because it corresponds to an initial-value problem with the reversed time axis. The secondary prototype may itself be unstable, but this certainly does not imply instability of the final SC network.

As an illustrative example we shall show that the excellent low sensitivity SC ladder filters (Lee *et al.*, 1981) may be synthesized by means of this method. Consider the doubly terminated LC reference circuit depicted in Fig. 8.1a. Its SSN-type equivalent circuit follows directly from the SSN-network transformation in section 6.2.4 and is presented in Fig. 8.1b. This circuit can be simplified and modified by omitting the sign inverters and by introducing the negative capacity

$$C_2' = -\frac{T^2}{4L_2}$$

as shown in Fig. 8.1c.

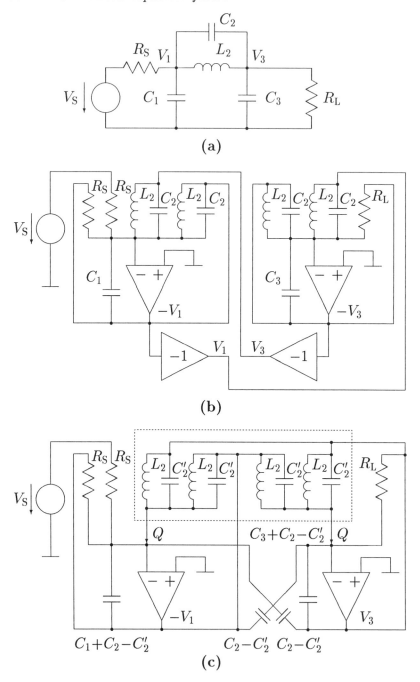

Fig. 8.1 Reference circuits for the illustrative example: (a) primary reference circuit, (b) SSN-type reference circuit, (c) modified SSN-type reference circuit.

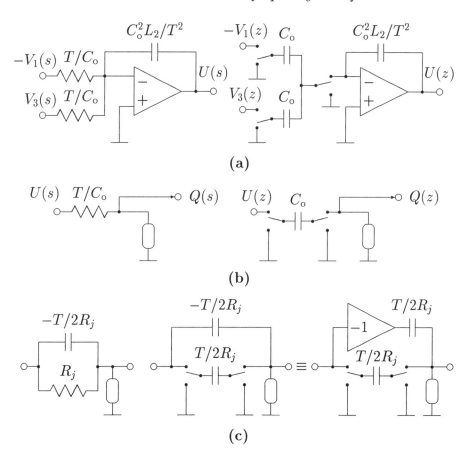

Fig. 8.2 Secondary reference subcircuits (prototypes) for: (a) transfer function $H_1(s)$, (b) transfer function $H_2(s)$, (c) resistor R_j.

Consider now the subcircuit indicated in Fig. 8.1c within the dotted box. It can be described by the transfer function

$$H(\psi) = \frac{Q(\psi)}{V_3(\psi) - V_1(\psi)}$$
$$= \frac{1}{\psi^2 L_2} + C_2' = \frac{1}{\psi^2 L_2} - \frac{T^2}{4L_2}$$
$$= \frac{T^2}{L_2}\left(\frac{2-\psi T}{2\psi T}\right)\left(\frac{2+\psi T}{2\psi T}\right). \qquad (8.8)$$

210   IIR switched-capacitor filters

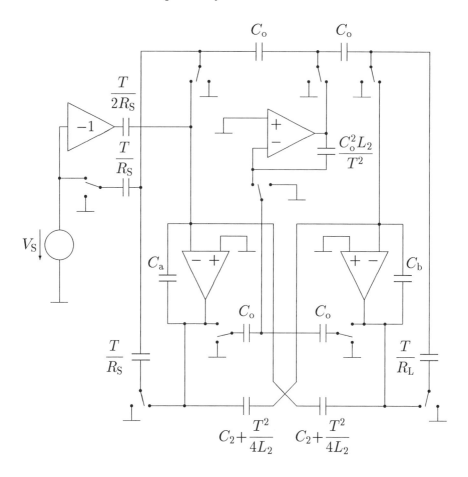

Fig. 8.3 Final SC realization.

Using the notation

$$(s_1 T)^{-1} = -\frac{2 - \psi T}{2\psi T} \quad \text{and} \quad (s_2 T)^{-1} = \frac{2 + \psi T}{2\psi T}$$

corresponding to the reversed forward Euler and backward Euler transformations, respectively, omitting (for simplicity) indices in variables s_1, s_2 and introducing an arbitrary capacitance C_o, equation (8.8) can be rewritten as

$$H(\psi) = H_1(s)H_2(s)$$

where

$$H_1(s) = \frac{U(s)}{V_3(s) - V_1(s)} = -\frac{T}{C_o L_2 s}$$

and

$$H_2(s) = \frac{Q(s)}{U(s)} = \frac{C_o}{sT}.$$

The secondary prototypes and their SC counterparts realizing transfer functions $H_1(s)$ and $H_2(s)$ are shown in Fig. 8.2a and b, respectively.

For the realization of resistors R_S and R_L we also use the backward Euler transformation. The secondary prototype and the corresponding SC schemes are shown in Fig. 8.2c.

The final SC realization is presented in Fig. 8.3, where the following simplifying denotations are introduced

$$C_a = C_1 + C_2 + \frac{T^2}{4L_2} - \frac{T}{2R_S}$$

and

$$C_b = C_2 + C_3 + \frac{T^2}{4L_2} - \frac{T}{2R_L}.$$

8.2 VIS-SC CIRCUITS

VIS-SC circuits, proposed in (Fettweis, 1979a, b), are switched-capacitor circuits comprising special building-blocks called **voltage inverter switches** (VISs). VIS-SC circuits constitute a separate class of switched-capacitor circuits that correspond bilinearly to a specific doubly resistively terminated lossless reference circuit. They are characterized by extremely low sensitivity and robust stability, if properly designed. Moreover, they are inherently multiphase and are suitable for efficient multirate applications (Dąbrowski, 1986a).

8.2.1 VIS-SC circuit concept

The idea of VIS-SC circuits is based on, among other things, an ingenious control of charge flow in the SC circuit, the structure of which reflects exactly, i.e. by a one-to-one element correspondence, the structure of the respective reference circuit. It is due to this fact that the potentially marvellous sensitivity and stability properties of doubly resistively terminated lossless (e.g. ladder) reference circuits are fully preserved in the resulting SC realizations.

The charge floats in a VIS-SC circuit in the form of separate pulses occurring once every sampling (or operation) period T. The operation cycle is usually composed of some, say M_c, clock pulses of duration T_c, i.e.

$$T = M_c T_c \tag{8.9}$$

The charge can flow only during a specific clock pulse (usually the last) within any sampling cycle. The other, auxiliary clock pulses play a secondary role (cf. section 8.2.2).

Such a particular charge flow is guaranteed by means of VISs, whose principle of operation is explained further on. They must certainly be inserted in all independent SC network meshes (or loops).

Consider any (a jth) capacitor C_j in the VIS-SC circuit. Since each charge pulse q_j that flows through capacitor C_j occurs in a specific operation period, both the voltage v_{bj} across this capacitor just before this pulse and the voltage v_{aj} just after this pulse may be assigned to the same discrete-time instant, say $t_n = t_0 + nT$, $0 \le t_0 < T$, within the considered sampling period, i.e.

$$q_j = q_j(t_n) \;,\quad v_{aj} = v_{aj}(t_n) \;\text{and}\; v_{bj} = v_{bj}(t_n) \;.$$

Moreover

$$q_j(t_n) = \bigl[v_{aj}(t_n) - v_{bj}(t_n)\bigr] C_j \;. \tag{8.10}$$

Defining the effective current

$$i_j(t_n) = \frac{q_j(t_n)}{T} \tag{8.11}$$

and the effective resistance

$$R_j = \frac{T}{2C_j} \;, \tag{8.12}$$

equation (8.10) may be rewritten in the form

$$\frac{v_{aj}(t_n) - v_{bj}(t_n)}{2} = R_j i_j(t_n) \ . \tag{8.13}$$

Comparing this result with equation (3.56a), we conclude that voltages $v_{aj}(t_n)$ and $v_{bj}(t_n)$ can be interpreted as the incident and the reflected wave, respectively.

Note that currents flow in a VIS-SC circuit according to the Kirchhoff current law and reflect respective relationships in the reference circuit. In order to achieve a similar correspondence for voltages following from the Kirchhoff voltage law, VISs should be voltage neutral, i.e. they should introduce no distortion to Kirchhoff voltage law equations. Define the voltage across capacitor C_j in instant t_n, in agreement with equation (3.56b), as

$$v_j(t_n) = \frac{v_{aj}(t_n) + v_{bj}(t_n)}{2} \ . \tag{8.14}$$

In order to achieve the voltage neutrality of VISs, they should model the short circuit. Consider for instance the kth VIS and denote its effective voltage in instant t_n as $v_k(t_n)$. Then, we postulate

$$v_k(t_n) = \frac{v_{ak}(t_n) + v_{bk}(t_n)}{2} = 0 \ ,$$

thus

$$v_{ak}(t_n) = -v_{bk}(t_n) \ . \tag{8.15}$$

Expression (8.15) describes the main function of a VIS and simultaneously explains the meaning of this term.

Now we shall consider realization of the reference circuit elements in the corresponding VIS-SC circuit. In order to realize all relevant lossless filters, it is enough to consider realization of resistive sources*, resistors, inductors and capacitors†.

---

*A resistive source (or real source) is a source with nonzero internal resistance, e.g. a voltage source connected in series with its internal resistance.

†Generalization for other important elements of lossless circuits, such as unit elements, circulators, gyrators, etc., is straightforward using their wave-flow models.

214 IIR switched-capacitor filters

Consider first resistor R_j in a reference circuit. Substituting relations (3.56a) and (3.56b) into equation (3.57a), we get the following condition

$$v_{bj}(t_n) = 0 \ . \tag{8.16a}$$

For inductor R_j we substitute relations (3.56a), (3.56b) into equation (3.57b) and obtain

$$v_{bj}(t_n) = -v_{aj}(t_{n-1}) \ . \tag{8.16b}$$

Similarly, but now using equation (3.57c), we obtain the following relation for capacitor R_j^{-1}

$$v_{bj}(t_n) = v_{aj}(t_{n-1}) \ . \tag{8.16c}$$

Finally, for a resistive source with voltage $e_j(t_n)$ and internal resistance R_j, we should guarantee

$$v_{bj}(t_n) = e_j(t_n) \ . \tag{8.16d}$$

Notice that all the above expressions constitute merely appropriate, simple initial conditions for a capacitor C_j. In accordance with equation (8.12), we can write

$$C_j = \frac{T}{2R_j} \ . \tag{8.17}$$

Hence, in the corresponding VIS-SC circuit, each reference circuit element is realized by capacitor C_j given by equation (8.17) and connected to a self-evident switch configuration fulfilling the appropriate initial condition: (8.16a)–(8.16d).

8.2.2 Realization of VIS elements

A VIS is a circuit element with one distinguished port. It operates cyclically with period T and in each operation cycle it inverts the voltage across its port. The VIS operation cycle usually consists of several phases numbered $1, 2, \ldots, i$. The first one is the registration phase, the last one is the voltage inversion phase, and the others are auxiliary phases. Between two consecutive operating cycles a

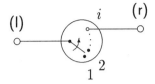

Fig. 8.4 Symbolic representation of a voltage inverter switch (VIS): 1 registration phase, $2, 3, \ldots, i-1$ auxiliary phases, i voltage inversion phase.

VIS is dormant, i.e. it carries no current. Since VISs are present in all independent circuit loops, the currents in VIS-SC filters flow as non-overlapping pulses occurring with period T. A general symbol used in this book for representing VISs is shown in Fig. 8.4.

As was shown by Fettweis (1979a, b; Fettweis, Herbst and Nossek, 1979), both grounded and floating VISs can be easily realized using op amps, capacitors and switches. Generally, two types of VISs may be distinguished:

- VISs based on the voltage registration (Fettweis, 1979a)

- VISs based on the principle of inverse recharging (Fettweis, 1979b).

In the first case, the VIS voltage v_b is registered and then the voltage $v_a = -v_b$ is forced across the VIS port. One of many possible configurations for a grounded voltage registration VIS is shown in Fig. 8.5a.

In the second case, the VIS is short-circuited during the registration phase and the charge flowing through its port is registered. Then this same charge is injected in the same direction, thus causing the voltage inversion. An illustrative example of a grounded VIS based on this principle is shown in Fig. 8.5b. Grounded VISs, insensitive to parasitic capacitances (which are unavoidable with present technology) can be designed using modified inverse recharging structures (Brückmann, 1986). VISs of this type are recommended for implementation in integrated circuits.

The principle of reverse recharging may also be used for the realization of floating VISs (Pandel and Dawei, 1981; Pandel, 1983b). Assume that the VIS in Fig. 8.4 is such a floating VIS. We consider

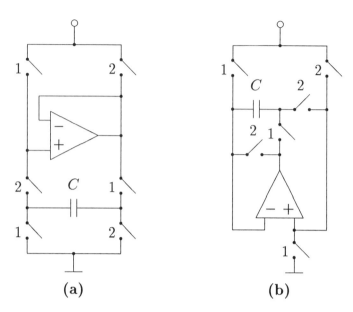

Fig. 8.5 Illustrative examples of grounded VISs: (a) VIS based on the voltage registration, (b) VIS based on the principle of inverse recharging.

its performance in some operation cycle. Denote VIS voltages from the left terminal (l) and the right terminal (r) to ground by $v^{(l)}$ and $v^{(r)}$, respectively. Let $v_b^{(l)}$, $v_b^{(r)}$ be the voltages just before the VIS voltage inversion begins and $v_a^{(l)}$, $v_a^{(r)}$ just after it ends. Moreover, let v_o be the voltage from both VIS terminals to ground when they are short-circuited. Then we can write

$$v_a^{(l)} - v_o = v_o - v_b^{(l)}$$

and

$$v_a^{(r)} - v_o = v_o - v_b^{(r)} \ .$$

Thus

$$v_a^{(l)} = 2v_o - v_b^{(l)} \tag{8.18a}$$

and

$$v_a^{(r)} = 2v_o - v_b^{(r)} \ . \tag{8.18b}$$

The floating VIS in Fig. 8.6 realizes equations (8.18a) and (8.18b). In phase 1, voltages $v_b^{(l)}$ and $v_b^{(r)}$ are registered at capacitors C_{1l} and

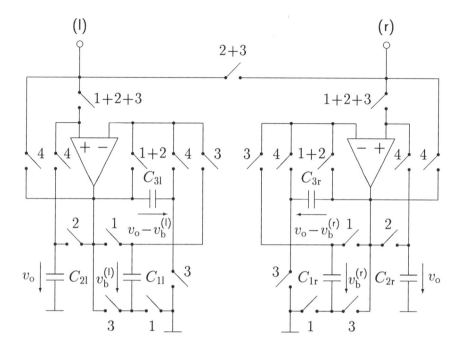

Fig. 8.6 Realization of a floating VIS based on the principle of inverse recharging.

C_{1r}, respectively. Then in phase 2, the VIS terminals are short-circuited and voltage v_o is stored in capacitors C_{2l} and C_{2r}. In an additional phase 3, the VIS terminals remain short-circuited and voltages $v_o - v_b^{(l)}$ and $v_o - v_b^{(r)}$ are stored in capacitors C_{3l} and C_{3r}, respectively. Finally, in the last (voltage inversion) phase 4, voltages $v_a^{(l)}$ and $v_a^{(r)}$ are computed according to equations (8.18a) and (8.18b), and forced at respective VIS terminals (voltages at capacitors C_{2l} and C_{2r} are added to voltages at capacitors C_{3l} and C_{3r}, respectively).

8.2.3 VIS-SC filter structures

VIS-SC filters can be designed using ladder structures (Herbst et al., 1979; Fettweis et al., 1980; Pandel, 1981; Pandel and Dawei, 1981; Herbst et al., 1982; Dąbrowski, 1988a) as well as lattice structures (Brückmann, 1984, 1986; Kleine, 1985).

218   *IIR switched-capacitor filters*

Ladder structures are often used for their excellent sensitivity. However, in order to eliminate the influence of bottom-plate parasitic capacitances, one is usually restricted to reference circuit structures with exclusively grounded elements. Such structures can be obtained by using unit elements. Their attenuation characteristics are, however, not as selective as those of LC filters of the same order. Therefore, VIS-SC lattice structures, which can be realized with grounded elements without degradation of their attenuation characteristics, may be advantageous in many applications. Moreover, lattice structures can be used to realize the multirate polyphase (two-channel) SBC scheme in Fig. 4.24. The main drawback of lattice filters is, however, their high sensitivity in the stopband with respect to element parameter variations. Hence, only VIS-SC filters, whose attenuation requirements for the stopband are not very high, can be realized by this method.

VIS-SC ladder arrangements are discussed in section 8.3. Therefore, now we concentrate our attention on VIS-SC lattice filters.

Notice first with regard to the design of VIS-SC lattice structures, no special technique for the realization of reflectances S_0 and S_1 (cf. equation (4.95) and Figs 4.29 and 4.30) is required. These reflectances are automatically realized by modelling the lattice impedances (reactances) Z_0 and Z_1 in Fig. 4.28 with VIS-SC branches. Although several structures can be used to realize reactances Z_0 and Z_1, unit element ladder structures with exclusively grounded elements seem to be the optimum choice.

Notice moreover that dual realizations may also be applied. It means that we can realize, for example, reactance Z_1^{-1} instead of reactance Z_1. Assume that constant R in equation (4.95) is equal to 1, then using Z_1^{-1}, we simply realize reflectance $-S_1$ instead of reflectance S_1. Now it is enough to replace equation (4.96a) by (4.96b) and vice versa.

As an illustrative example, a fifth-order lowpass VIS-SC lattice filter is shown in Fig. 8.7 (Brückmann, 1984). Reflectance S_0 is realized by the upper part of this circuit. It corresponds to reactance Z_0 that is synthesized as an unloaded chain of two unit elements. The lower part in turn realizes reflectance $-S_1$ corresponding to

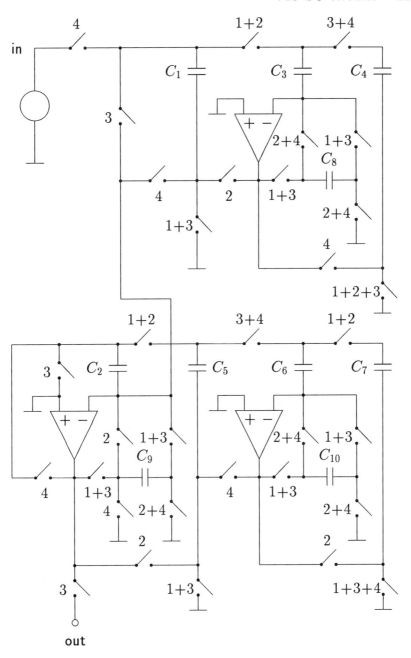

Fig. 8.7 An illustrative fifth-order lowpass VIS-SC lattice filter (Brückmann, 1984).

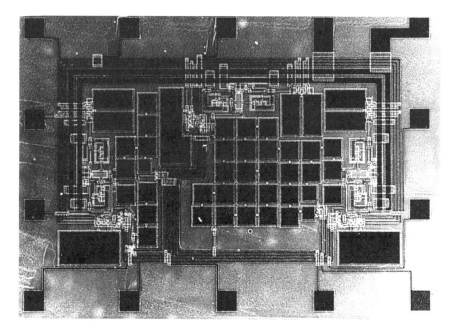

Fig. 8.8 Layout of the fifth-order lowpass VIS-SC lattice filter (Brückmann, 1984).

reactance Z_1^{-1}. This reactance is similarly synthesized but now the unloaded chain is composed of three unit elements. Capacitors C_1 and C_2 correspond to source resistance R. Capacitors C_3 and C_4 model unit element resistances of impedance Z_0, while capacitors C_5, C_6 and C_7 model unit elements of impedance Z_1^{-1}. Finally, capacitors C_8, C_9 and C_{10} are VIS capacitors realizing the inverse recharging. The input voltage is applied in phase 4 to capacitor C_1 but the same charge also flows through capacitor C_2, so the voltage across C_2 in this phase is equal $-C_1/C_2$ times the input voltage. This sign inversion eliminates the negative sign caused by realizing reflectance $-S_1$, so that equation (4.96b) should be still used for computing the lowpass output signal. In phase 3, we need only one VIS in the lower lattice branch. Capacitor C_9 is used in this phase to subtract the charges of capacitors C_1 and C_2, thus computing the output signal available as the op amp voltage at the end of phase 3. The layout of this filter is shown in Fig. 8.8.

Using sampled-and-held input signals, independent charging of capacitors C_1 and C_2, and an additional op amp, the VIS-SC lattice structure in Fig. 8.7 can be modified to implement the multirate branching filters of Figs 4.20 and 4.30. Thus, it can also be used to realize the two-channel SBC polyphase scheme in Fig. 4.24.

8.3 TRANSMISSION OF EFFECTIVE PSEUDOENERGY

8.3.1 Losses of effective pseudoenergy

In this section we shall consider multirate signal processing using the pseudolossless (paraunitary) systems discussed in section 3.2.4. We are particularly interested in the transmission of the effective pseudoenergy, i.e. the pseudoenergy of the so-called effective signal $x_e(nT)$ occurring at the input of the product modulator in the generalized multirate system in Fig. 4.7.

In the SBC (subband coding) schemes with QMF (quadrature-mirror filter banks) discussed in section 4.5.2, the whole input signal pseudoenergy is transmitted. It is split into different frequency bands in the analysis filter bank and then merged again in the synthesis filter bank. A completely different situation occurs if a multirate system (an interpolator, a decimator or, more generally, the generalized multirate system in Fig. 4.7) operates as a single block (Dąbrowski, 1989a, 1991). This is a quite practical situation occurring, for example, if a multirate approach is used to realize highly-selective filters (Franca and Mitra, 1994). Then losses of the effective pseudoenergy appear in the system of Fig. 4.7 and that in both filtering and product modulation. We shall consider these two kinds of losses separately.

Consider first the filtering losses. If the premodulation filter $H_1(e^{j\omega T})$ is ideal, then the premodulation filtering extracts merely the effective signal $x_e(nT)$. Thus, this process does not involve any losses of the effective pseudoenergy. Therefore, we can concentrate on the postmodulation filtering.

From condition (4.9) we immediately conclude that the total length of all passbands of transfer function $H_2(e^{j\omega T})$ equals $\Omega_o/2$ in the frequency band $\omega \in [0, \Omega/2]$.

The pseudoenergy of the signal $x_q(nT)$ can be expressed in the frequency domain by means of the spectrum $X_q(e^{j\omega T})$ using the Parseval equation (3.64a). On the basis of expression (4.15a), spectrum $X_q(e^{j\omega T})$ can have non-zero values for all frequencies in band $[0, \Omega/2]$. The idealized transfer function $H_2(e^{j\omega T})$, however, is non-zero only in its passbands of the total length $\Omega_o/2$, i.e. in $1/M$th of the band $[0, \Omega/2]$. Only this part of the pseudoenergy that is represented by the frequencies lying in the passbands of function $H_2(e^{j\omega T})$ is transmitted; the other part is lost.

Now we shall consider losses appearing in the product modulation. Denote by ϵ_{se} and ϵ_{sq} pseudoenergies of signals $x_e(nT)$, $x_q(nT)$, respectively, defined in accordance with expression (3.61). Assume that the carrier signal $q(nT)$ is composed of M-element modulation sequences $\tilde{q}$ and consider the polyphase decomposition of signal $x_e(nT)$ into M polyphase components $x_{e\nu}(nT)$, $\nu = 0, 1, \ldots, M-1$. Let $\epsilon_{se\nu}$ be the pseudoenergy and

$$\chi_{se\nu} = \frac{\epsilon_{se\nu}}{\epsilon_{se}} \qquad (8.19)$$

the relative pseudoenergy of the νth polyphase component. It is clear that

$$0 \leq \chi_{se\nu} \leq 1 \ , \quad \nu = 0, 1, \ldots, M-1 \ ,$$

and

$$\sum_{\nu=0}^{M-1} \chi_{se\nu} = 1 \ . \qquad (8.20)$$

Now we shall show that if signal $x_e(nT)$ corresponds to an integer L-band signal with respect to frequency $\Omega_o = 2\pi/(MT) = 2\pi/T_o$, then the pseudoenergies $\epsilon_{se\nu}$ of all its polyphase components $x_{e\nu}(nT)$ are equal, so equation (8.20) reduces to

$$\chi_{se\nu} = \frac{1}{M} \ , \quad \nu = 0, 1, \ldots, M-1 \ . \qquad (8.21)$$

In order to prove equation (8.21), notice from expression (3.51a)

that
$$X_{e\nu}(e^{j\omega T}) = \frac{1}{M}\sum_{m=0}^{M-1} e^{-jm\Omega_o \nu T} X_e(e^{j(\omega - m\Omega_o)T}) , \qquad (8.22)$$

$$\nu = 0, 1, \ldots, M-1 .$$

Now using the Parseval equation (3.64a), we get

$$\epsilon_{se} = \frac{2}{\Omega}\int_0^{\Omega/2} |X_e(e^{j\omega T})|^2 d\omega , \qquad (8.23a)$$

$$\epsilon_{se\nu} = \frac{2}{\Omega}\int_0^{\Omega/2} |X_{e\nu}(e^{j\omega T})|^2 d\omega . \qquad (8.23b)$$

Substituting equations (8.23a) and (8.23b) into (8.19) and using expression (8.22), we get

$$\chi_{se\nu} = \frac{\epsilon_{se\nu}}{\epsilon_{se}}$$

$$= \frac{1}{M^2} \frac{\int_0^{\Omega/2} \left| \sum_{m=0}^{M-1} e^{-jm\Omega_o \nu T} X_e(e^{j(\omega - m\Omega_o)T}) \right|^2 d\omega}{\int_0^{\Omega/2} |X_e(e^{j\omega T})|^2 d\omega} , \qquad (8.24)$$

$$\nu = 0, 1, \ldots, M-1 .$$

Since we assumed that signal $x_e(nT)$ corresponds to an integer L-band signal with respect to frequency Ω_o, spectra summed in the numerator of expression (8.24) do not overlap. Thus, finally

$$\chi_{se\nu} = \frac{1}{M^2} \frac{\sum_{m=0}^{M-1} \int_0^{\Omega/2} |X_e(e^{j\omega T})|^2 d\omega}{\int_0^{\Omega/2} |X_e(e^{j\omega T})|^2 d\omega} = \frac{1}{M} , \qquad (8.25)$$

$$\nu = 0, 1, \ldots, M-1 .$$

From equations (4.13), (4.16), (4.17) and (8.21), we conclude that

$$\epsilon_{sq} = \sum_{\nu=0}^{M-1} \epsilon_{se\nu} q^2(\nu T) = \epsilon_{se} \sum_{\nu=0}^{M-1} \chi_{se\nu} q^2(\nu T)$$

$$= \frac{\varepsilon_q}{M}\epsilon_{se} = P_q \epsilon_{se} . \qquad (8.26)$$

224   *IIR switched-capacitor filters*

For elementary modulation sequences with M_1 elements equal to ± 1 (the other being equal zero), equation (8.26) reduces to the form

$$\epsilon_{\text{sq}} = \frac{M_1}{M} \epsilon_{\text{se}} . \tag{8.27}$$

Further considerations are restricted to sampling rate alteration. In interpolation, pseudoenergy losses occur in postmodulation filtering only, since modulation is replaced, in this case, by sampling rate expansion, the process that does not change the signal pseudoenergy. On the other hand, in the case of decimation, only modulation losses appear (with parameter $M_1 = 1$), as the postmodulation filter does not exist in this case. Thus, in both processes, the interpolation and the decimation, only $1/M$th of the effective pseudoenergy is transmitted to the output.

A consequence of the considered losses is not only the decrease of the system dynamic range by about $10 \lg M$ dB, but also reduction of the stability margin for systems operating under looped conditions, e.g. if they are included in the loop of a communication link (Fettweis and Meerkötter, 1975, 1977; Dąbrowski, 1986b, 1988b).

8.3.2   VIS-SC circuits with recovery of effective pseudoenergy

(a)   Concept of pseudoenergy recovery

In previous section, losses of the effective pseudoenergy, i.e. the phenomenon occuring in the multirate system of Fig. 4.7, has been studied. Now the problem of recovery of this pseudoenergy will be considered.

The idea of recovery of the effective pseudoenergy was first proposed in the context of multirate wave digital arrangements (Fettweis and Nossek, 1982; Dąbrowski and Fettweis, 1986, 1987) and then applied also to VIS-SC interpolators and decimators (Dąbrowski, 1988a). Among the advantages obtained are:

- increased dynamic range

- better stability properties (increased stability margin under looped conditions)

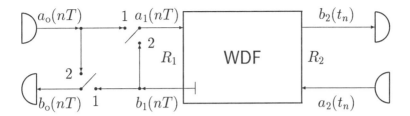

Fig. 8.9 Wave digital arrangement with recovery of effective pseudoenergy.

- improved filtering performance (greater stopband attenuation with the same filter structure).

The concept of recovery of effective pseudoenergy can be explained using the wave digital arrangement in Fig. 8.9. Such an arrangement can be applied to the premodulation filter (e.g. in a decimator) as well as to the postmodulation filter (e.g. in an interpolator). This system will operate as a postmodulation filter if $a_2(t_n) \equiv 0$, $a_0(nT)$ is the input signal and $b_2(t_n)$ is chosen to be the output signal. Conversely, the system will operate as a premodulation filter if $a_0(nT) \equiv 0$, $a_2(t_n)$ is the input signal and if $b_0(nT)$ is chosen to be the output signal.

The switches in Fig. 8.9 are in position 1 for those instants nT for which modulation signal samples $q(nT)$ are not zero ($q(nT) \neq 0$). For other instants, i.e. if $q(nT) = 0$, the switches are in position 2. By this means, a feedback loop occurs in switch position 2 at the left port of the wave digital filter (WDF) in Fig. 8.9. The effective pseudoenergy that is to be recovered is represented by the reflected signal $b_1(nT)$ that is transmitted back to the filter input $a_1(nT)$ if the switch is in position 2. Thus, the pseudopower $b_1^2(nT)/R_1$ is retrieved in every instant corresponding to switch position 2, by feeding it back to the filter. The filter converts this pseudopower into the appropriate frequency range and transmits it to the output in appropriate instants.

The above method of recovery of the effective pseudoenergy can also be illustrated using the reference filter concept. The appropriate circuits are shown in Fig. 8.10. The difference between these

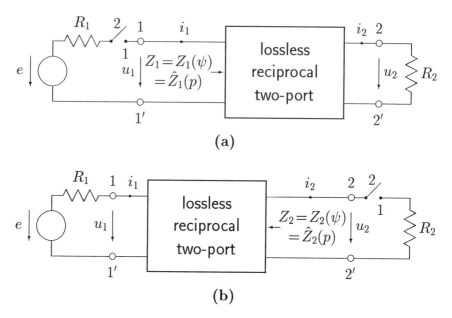

Fig. 8.10 Reference circuits corresponding to the multirate arrangements with recovery of the effective pseudoenergy: (a) for interpolation, (b) for decimation.

and the common reference circuit (i.e. that without pseudoenergy recovery) lies in the switch configuration. In the circuits of Fig. 8.10, for switch position 2, i.e. in instants of unavailable/unnecessary samples, the appropriate port is opened. Thus, we avoid in these instants the pseudoenergy losses at this port.

Now we attempt to adopt this idea for SC filters. For wave-SC filters, which simulate the wave-flow diagrams of WDFs (Kleine *et al.*, 1981; Mavor *et al.*, 1981), the problem of pseudoenergy recovery can be solved in just the same way as for WDFs, i.e. by feeding the signal $b_1(nT)$ back to the filter (Fig. 8.9). However, wave-SC filters require more complicated hardware than the equivalent VIS-SC filters and, therefore, they will not be discussed further.

For VIS-SC filters, this method of recovery of the effective pseudoenergy cannot be directly applied because signals $a_1(nT)$ and $b_1(nT)$ are not explicitly modelled in filters of this type. This is because these signals correspond to a filter port and not to a specific filter element. However, the desirable structures of VIS-SC filters with

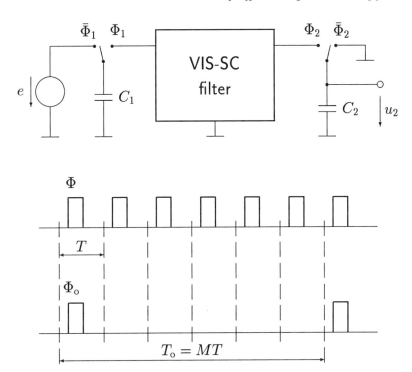

Fig. 8.11 VIS-SC filter based on the inverse recharging principle: for $\Phi_1 = \Phi_2 = \Phi$ this is a classical filter, for $\Phi_1 = \Phi_o$ and $\Phi_2 = \Phi$ this is an interpolator with recovery of the effective pseudoenergy, and for $\Phi_1 = \Phi$ and $\Phi_2 = \Phi_o$ this is a decimator with recovery of the effective pseudoenergy.

recovery of the effective pseudoenergy can be derived directly from the reference circuit of Fig. 8.10. One of the possible structures, namely that based on the inverse recharging principle, is shown in Fig. 8.11. In this case, the additional switch (corresponding to the switch in Figs 8.10a and b) is not necessary because its role can be assumed by the appropriate switch already existing in the filter, i.e. the input switch for interpolation and the output switch for decimation (cf. Fig. 8.11). Hence, realization of multirate VIS-SC filters with recovery of the effective pseudoenergy needs no change in the hardware (in comparison with classical VIS-SC filters) except for the modification of one of the switch controlling signals, as shown in Fig. 8.11, and for the change of values of the filter capacitors.

The new capacitances are to be found by optimization of the filter frequency characteristics (i.e. the conversion functions introduced in section 4.4.2).

It should be noticed that capacitors C_1 and C_2 in Fig. 8.11 are not connected with VISs. Elimination of these two VISs and the corresponding savings in the hardware are possible because, first, we do not need the explicit value of the input reflected signal u_{1a} and, second, we can, owing to the principle of inverse recharging, use voltage

$$u_2 = \frac{u_{2a}}{2}$$

as the filter output, thus eliminating the necessity of forming the voltage u_{2a}.

(b) Analysis of SC circuits with pseudoenergy recovery

In order to demonstrate the advantages of the VIS-SC circuit in Fig. 8.11, analysis concentrated on sampling rate alteration will be presented below. General analysis, i.e. that for an arbitrary modulation sequence, is much more complicated (Dąbrowski, 1988b).

Assume that the source signal for interpolation is given by

$$e(\tau_{1\nu}) = E e^{p\tau_{1\nu}}, \qquad (8.28a)$$

$$\tau_{1\nu} = \tau_{10} + \nu T_o, \quad T_o = MT, \quad 0 \leq \tau_{10} < T_o, \quad \nu = 0, \pm 1, \pm 2, \ldots,$$

while that for decimation is given by

$$e(t_{1n}) = E e^{p t_{1n}}, \qquad (8.28b)$$

$$t_{1n} = t_{10} + nT, \quad 0 \leq t_{10} < T, \quad n = 0, \pm 1, \pm 2, \ldots,$$

where E and p are complex constants.

To find expressions for the output signal u_2, notice that the circuit under consideration is linear but periodically time-varying with period T_o. Thus, it can be analysed using the approach presented in section 4.4.

Consider first interpolation. Adopting expression (4.27b) and substituting this in (4.40), we can write for a steady-state in the circuit

$$u_2(t_{2n}) = e^{pt_{2n}} \sum_{m=0}^{M-1} U_{2m} e^{jm\Omega_o t_{2n}}, \qquad (8.29a)$$

$$t_{2n} = t_{20} + nT, \quad 0 \le t_{20} < T, \quad n = 0, \pm 1, \pm 2, \ldots.$$

Following the notation in equation (4.40), we can write

$$U_{2m} = U_{2m}\left(e^{j(\omega + m\Omega_o)T}\right).$$

Complex values U_{2m}, $m = 0, 1, \ldots, M-1$, are proportional to the complex constant E.

On the other hand, for decimation we obtain

$$u_2(\tau_{2\nu}) = U_2 e^{p\tau_{2\nu}}, \qquad (8.29b)$$

$$\tau_{2\nu} = \tau_{20} + \nu T_o, \quad T_o = MT, \quad 0 \le \tau_{20} < T_o, \quad \nu = 0, \pm 1, \pm 2, \ldots,$$

where

$$U_2 = U_2\left(e^{j\omega T}\right).$$

From expression (8.29a), it follows that the interpolator may be described by M conversion functions

$$K_{21m} = \frac{U_{2m}}{E}, \quad m = 0, 1, \ldots, M-1, \qquad (8.30a)$$

while the decimator is described by only one conversion function

$$K_{21} = \frac{U_2}{E}. \qquad (8.30b)$$

Following the pseudopower concept (section 3.2.3), it is useful to multiply the 'natural' conversion functions (8.30a) and (8.30b) by a positive constant such that the magnitudes of the resulting 'normalized' conversion functions are equal to the square roots of the corresponding average pseudopowers. Thus, we define the normalized conversion functions as

$$H_{21m} = 2\sqrt{\frac{MC_2}{C_1}} K_{21m}, \quad m = 0, 1, \ldots, M-1, \qquad (8.31a)$$

for the interpolator and

$$H_{21} = 2\sqrt{\frac{C_2}{MC_1}} K_{21} \tag{8.31b}$$

for the decimator.

Dąbrowski and Fettweis (1986, 1987) analysed the recovery of effective pseudoenergy in multirate wave digital arrangements and stated the normalized conversion functions as

$$\begin{aligned}H_{21m} &= \frac{2\sqrt{MR_1R_2}\hat{M}_\mathrm{I}(p+jm\Omega_\mathrm{o})}{\tilde{Z}_1 + MR_1} \\ &= \frac{2\sqrt{MR_1R_2}}{\tilde{Z}_1 + MR_1} M_\mathrm{I}\left(\frac{\psi + j\tan(m\pi/M)}{1 + j\psi\tan(m\pi/M)}\right), \quad (8.32\mathrm{a})\\ &m = 0,1,\ldots,M-1,\end{aligned}$$

and

$$H_{21} = \frac{2\sqrt{MR_1R_2}M_\mathrm{U}(\psi)}{\tilde{Z}_2 + MR_2} \tag{8.32b}$$

where variables p and ψ are related by equation (3.54), $M_\mathrm{I} = M_\mathrm{I}(\psi) = \hat{M}_\mathrm{I}(p)$ is the current transfer function from port 1 to port 2 of the reference circuit in Fig. 8.10a, $M_\mathrm{U} = M_\mathrm{U}(\psi) = \hat{M}_\mathrm{U}(p)$ is the open-circuit voltage transfer ratio from port 1 to port 2 of the reference circuit in Fig. 8.10b (R_2 is replaced by an open-circuit), and $\tilde{Z}_1$, $\tilde{Z}_2$ are the discrete pulse impedances defined as

$$\tilde{Z}_j = \sum_{m=0}^{M-1} \hat{Z}_j(p + jm\Omega_\mathrm{o}), \quad j = 1,2. \tag{8.33}$$

Assume for simplicity of further considerations that the circuit in Fig. 8.10b contains the same reciprocal two-port as in Fig. 8.10a but reflected with respect to the left- and right-hand sides, then the following relations hold

$$M_\mathrm{I} = M_\mathrm{U} = M_{21}(\psi) \tag{8.34}$$

and

$$H_{210} = H_{21}. \tag{8.35}$$

Following the concept of fitted conversion functions T_m and the resultant conversion function T, $m = 0,1,\ldots,M-1$, presented in section 4.4.2, we conclude that it is possible to determine all transmission properties of the systems under consideration by only one function, the so-called **resultant conversion function**

$$T = T(\psi) = T_0(\psi) = \ldots = T_{M-1}(\psi) = H_{210}(\psi) = H_{21}(\psi) \ . \quad (8.36)$$

Indeed, since $\tilde{Z}_1$ is periodic in p with period $j\Omega_o$, the relation

$$\tilde{T}_m(p + jm\Omega_o) = \tilde{H}_{21m}(p) = \tilde{H}_{210}(p + jm\Omega_o) \ , \quad (8.37)$$

$$m = 0, 1, \ldots, M - 1 \ ,$$

holds.

From expressions (8.32a) and (8.32b), it follows that using quantities normalized with respect to R_2, the resultant conversion function can be expressed as

$$T = T(\psi) = \frac{2\sqrt{r_1} M_{21}(\psi)}{\tilde{z}_1 + r_1} \quad (8.38)$$

where

$$r_1 = \frac{MR_1}{R_2} \quad (8.39a)$$

and

$$\tilde{z}_1 = \frac{\tilde{Z}_1}{R_2} \ . \quad (8.39b)$$

Dąbrowski and Fettweis (1987) elaborated the realizability conditions for the resultant conversion function $T = T(\psi)$ under assumptions that the reference circuits to the systems under consideration are lossless, reciprocal and composed exclusively of elements described by rational functions of the complex frequency ψ. The last assumption seems to be too strong for VIS-SC filters. This is because we would like to let unit elements (Fraiture and Neirynck, 1969) be present in the reference circuits of these filters. Therefore, the theory given by Dąbrowski and Fettweis (1987) must be modified for multirate VIS-SC filters with pseudoenergy recovery, as is shown below (Dąbrowski, 1988a).

We can write the normalized input impedance (Fig. 8.10a)

$$z_1(\psi) = \frac{Z_1}{R_2}$$

in the form

$$z_1 = z_1(\psi) = k_{-1} + k_0 + \sum_{i=1}^{\kappa} \frac{k_i}{\psi - \psi_i} \qquad (8.40)$$

where k_{-1}, k_0 are real nonnegative constants and k_i, ψ_i are in general complex constants occuring, if not real, in conjugate pairs. Moreover, $\Re\psi_i \leq 0$ and if $\Re\psi_i = 0$ then k_i is real and positive. We may always assume that

$$\Re\psi_i \begin{cases} < 0 & \text{for } i = 1, 2, \ldots, \nu < \kappa \ , \\ = 0 & \text{for } i = \nu + 1, \nu + 2, \ldots, \kappa \ . \end{cases} \qquad (8.41)$$

Additionally, we assume that the two-ports in Figs 8.10a and b contain, in addition to lumped elements, unit elements occuring in configurations as usually encountered (Fraiture and Neirynck, 1969). Then, the transfer ratio $M_{21} = M_{21}(\psi)$ can be put into the form

$$M_{21}(\psi) = \frac{f}{d} \qquad (8.42)$$

where $d = d(\psi)$ is a real Hurwitz polynomial and $f = f(\psi)$ is a real, even or odd quasi-polynomial of the form

$$f = f(\psi) = f_1(\psi) \left(1 - \psi^2\right)^{\mu_2/2} \qquad (8.43)$$

where $f_1 = f_1(\psi)$ is a real, even or odd polynomial of degree μ_1.

Assume that the polynomial $d = d(\psi)$, obtained as stated below, has only simple zeros. In this case we say that the two-ports under consideration are strictly normal. Thus, for the degree ν of d we have

$$\nu \geq \mu_1 + \mu_2 \qquad (8.44)$$

and

$$d = d(\psi) = \prod_{i=1}^{\nu} (\psi - \psi_i) \ . \qquad (8.45)$$

Moreover,

$$k_0 = f_{1(\nu-\mu_2)}^2 \qquad (8.46a)$$

where $f_{1(\nu-\mu_2)}$ is the coefficient by term $\psi^{\nu-\mu_2}$ of polynomial $f_1 = f_1(\psi)$ and

$$k_i = \frac{(-1)^{\mu_1-1} f^2(\psi_i)}{\psi_i \prod_{\substack{l=1 \\ l \neq i}}^{\nu}(\psi_l^2 - \psi_i^2)} = \frac{2(-1)^{\mu_1} f^2(\psi_i)}{d'(\psi_i) d(-\psi_i)}$$

$$= \frac{2(-1)^{\mu_1} f^2(\psi_i)}{c'(\psi_i)}, \qquad (8.46b)$$

$$i = 1, 2, \ldots, \nu,$$

where the polynomial $c = c(\psi) = d(\psi)d(-\psi)$. The other parameters in equation (8.40), i.e. k_{-1} and k_i, $i = \nu+1, \nu+2, \ldots, \kappa$, are arbitrary real nonnegative constants subject only to the restriction that to two conjugate ψ_i there correspond k_i that are equal.

(c) Filter optimization

If a multirate VIS-SC filter with recovery of the effective pseudoenergy were built with capacitances of the classical filter (i.e. of that without pseudoenergy recovery), then the filter attenuation characteristic would be unsatisfactory (c.f. illustrative example below). This is why optimization of the capacitances is necessary. The aim is to obtain the attenuation characteristic which, on the one hand, meets the passband requirements, and, on the other hand, yields the maximum possible stopband loss.

We discretize the frequency axis and define the following quantities

$$\alpha_{p\,\max}^{(q)} = \alpha_{p\,\max} \left[\sum_{\substack{k \\ \text{pass-} \\ \text{band}}} \left(\frac{\alpha_{pk}}{\alpha_{p\,\max}} \right)^q \right]^{1/q}, \qquad (8.47a)$$

$$\alpha_{p\,\min}^{(q)} = \alpha_{p\,\min} \left[\sum_{\substack{k \\ \text{pass-} \\ \text{band}}} \left(\frac{\alpha_{pk}}{\alpha_{p\,\min}} \right)^{-q} \right]^{-1/q}, \qquad (8.47b)$$

234 IIR switched-capacitor filters

$$\alpha_{s\,min}^{(q)} = \alpha_{s\,min} \left[\sum_{\substack{l \\ \text{stop-} \\ \text{band}}} \left(\frac{\alpha_{sl}}{\alpha_{s\,min}} \right)^{-q} \right]^{-1/q}, \quad (8.47c)$$

where α_{pk} is the attenuation at the kth frequency point in the passband, $\alpha_{p\,max}$ is the maximum attenuation in the passband, $\alpha_{p\,min}$ is the minimum attenuation in the passband, α_{sl} is the attenuation at the lth frequency point in the stopband, $\alpha_{s\,min}$ is the minimum attenuation in the stopband, and q is a positive parameter.

We define the objective function which is to be minimized as

$$F = \gamma \left(\frac{\alpha_p^{(q)}}{\alpha_{p\,ref}} - 1 \right)^2 - \alpha_s^{(q)} \quad (8.48)$$

where $\alpha_{p\,ref}$ is the reference attenuation ripple in the passband, γ is the weight factor, parameter $\alpha_s^{(q)}$ is defined by

$$\alpha_s^{(q)} = \alpha_{s\,min}^{(q)} - \alpha_{p\,max}^{(q)} \quad (8.49)$$

and parameter $\alpha_p^{(q)}$ is defined according to two different optimization strategies:

- **Strategy 1**: optimization with minimum passband flat loss (minimum reflected pseudopower)

$$\alpha_p^{(q)} = \alpha_{p\,max}^{(q)} \quad (8.50a)$$

where $\alpha_{p\,max}^{(q)}$ is defined in accordance with the attenuation normalized with respect to the incident pseudopower,

- **Strategy 2**: optimization with arbitrary passband flat loss

$$\alpha_p^{(q)} = \alpha_{p\,max}^{(q)} - \alpha_{p\,min}^{(q)}. \quad (8.50b)$$

As an illustrative example of the presented theory and optimization strategies, the design, optimization and measurement of a fifth-order low-pass VIS-SC filter for a sampling rate increase of a factor $N = 2$ is presented below. The filter was designed to have a passband edge at 3.415 kHz, a stopband edge at 4.584 kHz and a passband ripple of at most 0.04365 dB.

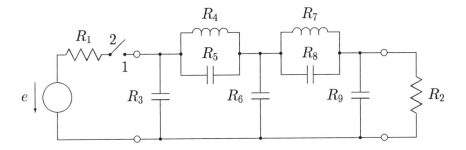

Fig. 8.12 Reference circuit for the illustrative two-fold fifth-order interpolator.

The VIS-SC filter was designed using the ladder reference circuit of Fig. 8.12. The floating VISs of Fig. 8.6, based on the principle of inverse recharging, were aplied. The resulting VIS-SC circuit scheme together with the final SC realization are shown in Figs 8.13a and b, respectively. In Figs 8.13c and d clock diagrams are shown for the classical interpolator (i.e. that without recovery of the effective pseudoenergy) and for the interpolator with pseudoenergy recovery, respectively. For the former, filter capacitances corresponding to element values of Cauer filter C051039 (Saal, 1979) were used. For the latter, capacitances were optimized by both above optimization strategies, resulting in two circuit versions.

In order to demonstrate the advantages of the presented method for recovery of the effective pseudoenergy, computer simulations and measurements were carried out. The normalized ideal characteristics (i.e. those computed assuming ideal op amps and neglecting parasites) are shown in Fig. 8.14 for Cauer filter C051039 and for three filters with pseudoenergy recovery: a filter with the original capacitances (those of the Cauer filter), a filter with capacitances optimized by Strategy 1, and a filter with capacitances optimized by Strategy 2. The unsatisfactory passband waveform of the characteristic in Fig. 8.14b for the filter with pseudoenergy recovery but with the original capacitances (the maximum passband ripple being ca. 0.6 dB, i.e. about 14 times larger than permissible) illustrates the necessity of the optimization of filter capacitances. The minimum stopband attenuations of the optimized filters are greater by

236 IIR switched-capacitor filters

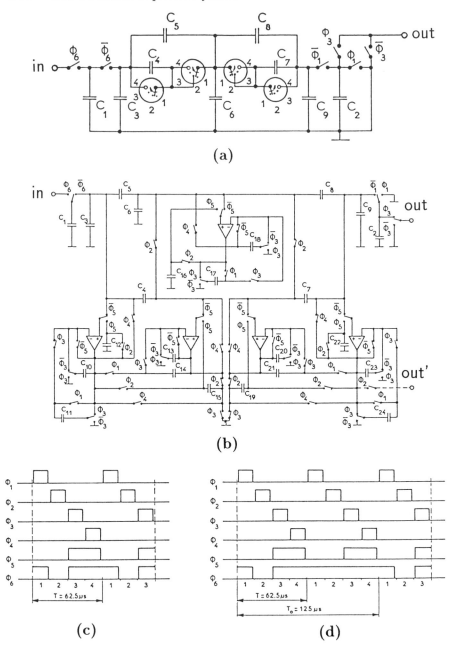

Fig. 8.13 VIS-SC realization of the illustrative two-fold fifth-order interpolator: (a) simplified SC circuit scheme, (b) detailed SC circuit scheme, (c) classical clock diagram, (d) clock diagram for pseudoenergy recovery.

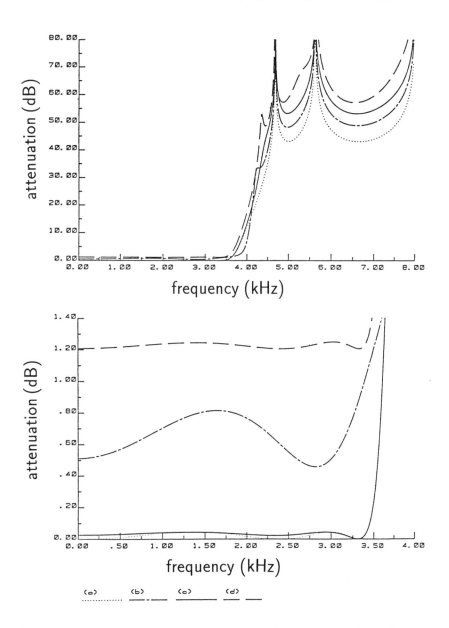

Fig. 8.14 Theoretical attenuation characteristics for four interpolator filter versions: (a) Cauer filter C051039, (b), (c), (d) filters with pseudoenergy recovery: (b) with original capacitances, (c) with capacitances optimized by Strategy 1 and (d) with capacitances optimized by Strategy 2.

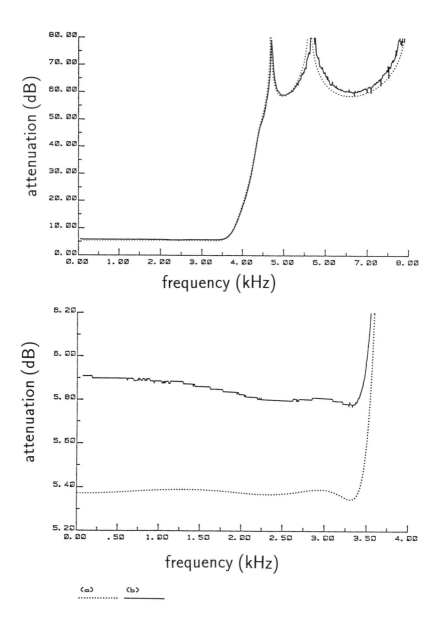

Fig. 8.15 Comparison of the computed and measured attenuation characteristic for the illustrative interpolator filter with capacitances optimized by Strategy 1: (a) computed curve, (b) measured curve.

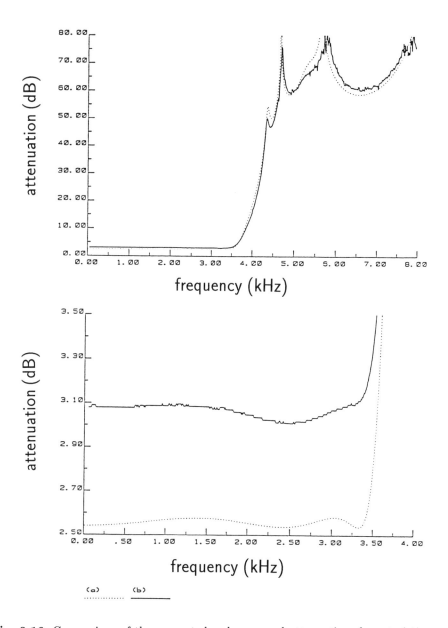

Fig. 8.16 Comparison of the computed and measured attenuation characteristic for the illustrative interpolator filter with capacitances optimized by Strategy 2: (a) computed curve, (b) measured curve.

10.2 dB for Strategy 1 and by 13.0 dB for Strategy 2 than the minimum stopband attenuation of the Cauer filter. In order to show that only quite small deviations occur between the measured and computed attenuation characteristics, the respective curves are compared in Fig. 8.15 for the filter optimized by Strategy 1 and in Fig. 8.16 for the filter optimized by Strategy 2. The excellent insensitivity of these filters with respect to capacitor values and to parasites has also been proven by measurements.

Finally, it should be stressed that all advantages of the recovery of the effective pseudoenergy are achieved in practice without any additional cost, since new filters are realized by the same hardware as their classical Cauer counterpart. The only changes are the new (optimized) capacitor values and one modified clock signal.

9

Applications of multirate and multiphase SC circuits

In this chapter a short overview of the applications of multirate and multiphase switched-capacitor circuits with emphasis towards multirate analog/digital signal processing is presented. Among many diverse, important applications, the following have been chosen and are briefly discussed below: transmultiplexers, highly-selective filters, data converters and adaptive filters.

9.1 TRANSMULTIPLEXERS AND HIGHLY-SELECTIVE FILTERS

9.1.1 Single-way transmultiplexers

One of the most interesting transmultiplexer (TDM/FDM or audio/FDM converter) concepts is that proposed by Fettweis (1978a, 1978b). It is based on the so-called **multistage single-way** schemes. Such transmultiplexers require product modulators with multiplications by only $+1$, -1 and 0 (cf. Table 4.1). Not only for this reason but also because of avoiding the balancing problem, due to the single-way transmission, these systems can successfully be implemented using switched-capacitor techniques, e.g. the VIS-SC filters discussed in sections 8.2 and 8.3 (Fettweis, 1982; Dąbrowski, 1989b). Especially efficient are the periodically time-varying schemes with recovery of effective pseudoenergy presented in section 8.3.

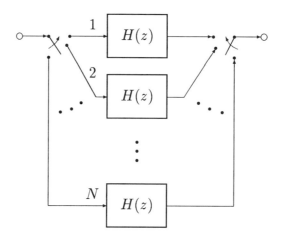

Fig. 9.1 N-path filtering structure.

9.1.2 N-path and pseudo-N-path filters

Bandpass filters required, for example, in transmultiplexer schemes could certainly be realized on the basis of appropriate bandpass reference circuits. This results, however, in an increased circuit sensitivity and in an unacceptably large capacitance spread, especially for highly selective filters. A much better approach lies in the application of the N-path filtering principle (Franks and Sandberg, 1960; Grünigen et al., 1981), which can also be combined with the frequency translation technique realized by means of multirate signal processing. This technique is discussed in next section.

An N-path discrete-time filter is composed of N identical path filters (each of them operating with, say, rate F) switched in such a way that they are sequentially connected between the input and the output terminals, thus realizing their sampling with an overall rate NF (Fig. 9.1). In consequence, the resulting N-path filter has the same frequency characteristic as each of its path filters* but the Nyquist frequency $NF/2$ is now N times larger. The simplest

---

*This is only true if sample-and-hold effects at the output are omitted. In reality, for the sampled-and-held output signal, the ideal frequency characteristic must be multiplied by function $\sin(\pi f/F)/(\pi f/F)$ for a single-path filter or by $\sin(\pi f/NF)/(\pi f/NF)$ for an N-path filter.

realization of a bandpass filter of this type is a two-path filter with high-pass path filters.

The main drawback of analog, i.e. also SC, realizations based on the N-path filtering principle is their big sensitivity to the path mismatch. Therefore, the so-called pseudo-N-path filters have also been proposed (Fettweis and Wupper, 1971). This modified approach consists in the realization of the N paths in such a way that each delay of the sampled-data prototype (i.e. a single-path filter) is replaced by an N-stage shift register. It means that in such filters only one path physically exists but is multiplexed by time sharing. The advantages of such filters, in comparison with their real N-path counterparts, are

- simplified hardware

- practically ideal balancing of the paths.

These features are very important for SC realizations. Therefore the pseudo-N-path filtering principle has been readily adopted for them (Pandel, 1982, 1983a; Pandel *et al.*, 1986); combining this with the recovery of the effective pseudoenergy leads to realizations with greater dynamic range and better attenuation characteristics (Dąbrowski, 1989b).

9.1.3 Frequency translated SC filters

As already mentioned in section 3.1.3, an SC filter, as a discrete-time signal-processing system, should be preceded by an anti-aliasing filter (AAF) and followed by an anti-imaging filter (AIF). Both these filters should allow to pass two, in general different, integer L-band signals with respect to the SC filter operating frequency $\Omega = 2\pi F$. Freedom in the choice of passbands of these two filters results in a tremendous variety of combinations that lie far beyond the most typical situation where both AAF and AIF filters are low-pass filters with the passband cut-off frequency equal to the Nyquist frequency $F/2$. This kind of technique can be used to translate the signal frequency in SC filters.

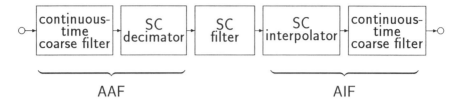

Fig. 9.2 General highly selective multirate band-pass filtering architecture.

A quite practical application of the frequency translation technique is the design of highly selective bandpass filters, i.e. filters with a very narrow relative bandwidth, below, say 1%, and with a very high Q-factor (cf. Fig. 1.5). The multirate approach can be combined with this technique (Franca and Haigh, 1988), resulting in a general bandpass filtering architecture of this type, shown in Fig. 9.2. An AAF is realized in this case as a cascade of a relative simple and coarse continuous-time filter followed by an SC bandpass decimator. By this means the appropriate frequency-translated band (certainly towards low frequencies) is selected for the actual, precise, low-frequency SC bandpass filter. Then, the SC bandpass interpolator followed by another continuous-time, coarse filter together form an AIF that performs the complementary operation and selects the appropriate frequency band for the output signal. The interpolation process is somehow disturbed through additional attenuation caused by the sampled-and-held nature of the SC bandpass filter output signal.

9.2 DATA CONVERTERS

Transition from analogue to digital data and vice versa (cf. sections 3.1.1 and 3.1.2) requires analog-to-digital (A/D) and digital-to-analog (D/A) data converters, respectively. In the past, analog–digital interfaces were developed and manufactured as separate systems or built with individual integrated circuits. Nowadays, modern technologies make the fabrication of mixed analog–digital ASICs (application specific integrated circuits) possible. Such circuits can realize very complex systems comprising, for example, a large number of data converters combined with quite complicated digital signal processing in a single chip.

One of the most interesting applications of multirate and multiphase switched-capacitor circuits is in systems of this kind. Among various approaches to data conversion, oversampling techniques, polyphase decomposition and QMF filtering play a very important role.

9.2.1 Oversampling converters

There are basically two ways for the realization of precise data converters:

- Circuits operating with low sampling rate (forming the so-called Nyquist rate class of data converters). Such circuits require very accurate analog components, and thus special analog technologies are necessary for their fabrication. Moreover, additional, complicated analog continuous-time anti-aliasing filters are required.

- Circuits operating with a much higher than Nyquist sampling rate, i.e. the so-called **oversampling converters**. Such systems are usually characterized by a very low (e.g. one bit only) resolution which is then suitably increased by decimation. Such circuits can be realized using standard digital technologies. They are therefore required in modern mixed analog–digital integrated circuits.

Among techniques used in oversampling converters, pulse width modulation (PWM) and deterministic or stochastic pulse density modulation (PDM) can be distinguished. The wide spectrum quantization noise due to oversampling (cf. section 3.1.2) is then filtered out in the process of decimation, thus increasing the signal resolution measured as the number of wordlength bits of the corresponding digital signal.

A better technique consists in an additional noise shaping so that the noise component in the interesting Nyquist range $(0, F/2)$

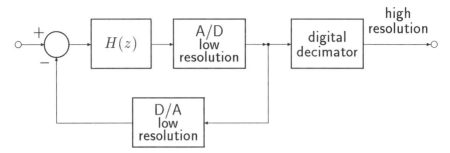

Fig. 9.3 Noise-shaping feedback loop in an oversampling A/D converter.

is significantly reduced. It can be realized using a special low-resolution A/D–D/A feedback loop (Maloberti, 1991). This method is illustrated in Fig. 9.3.

Circuits with a resolution of only one bit are commonly used in the noise-shaping feedback loops. They are known as **sigma-delta modulators** (Boser and Wooley, 1988).

9.2.2 Polyphase converters

A special class of high speed A/D converters is constituted by multirate circuits based on the idea of polyphase decomposition (sections 3.1.6 and 4.4.4), possibly combined with subband decomposition using QMF filter banks (sections 4.5.1 and 4.5.2). This concept consists in the realization of a high-speed A/D converter using a number of low-speed and low-cost A/D subconverters interleaved in time as indicated in Fig. 9.4 (Petraglia and Mitra, 1992, 1993). An analysis filter bank is realized as an analog circuit using multirate switched-capacitor techniques. The synthesis filter bank is digital. Such converters are suitable for applications of up to about 100 MHz, i.e. for video range applications.

Among the advantages thus obtainable are all good points of the subband coding discussed in section 4.5, savings in the chip area and a reduction in power consumption. Potential drawbacks, however, are sampling jitter and aliasing distortion.

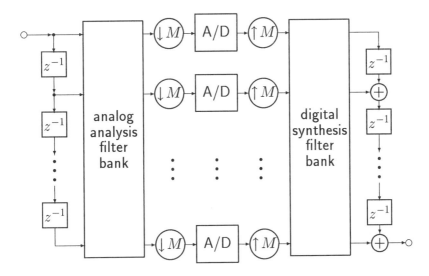

Fig. 9.4 Subband decomposition A/D converter.

9.3 ADAPTIVE FILTERS

Adaptive filtering techniques have gained widespread acceptance in many applications of signal processing. Typically, adaptive filters are implemented as digital transversal FIR filters and the LMS (**least-mean-square**) adaptation algorithm is used to adjust the zeros of the adaptive filter (e.g. Treichler, 1987). However, for certain (in particular low-power) applications, analog filters, and especially switched-capacitor adaptive filters, have distinct advantages over their all-digital counterparts. Adaptation speed can be significantly increased using subband and polyphase decomposition, i.e. multirate techniques (Somayazulu, Mitra and Shynk, 1989).

On the other hand, the order of an adaptive filter can, in many cases, be significantly reduced through the use of an IIR filter (Shynk, 1989; Dąbrowski and Braszak, 1992), where either both the poles and the zeros or merely the zeros are adapted. Therefore recently SC integrated circuits for the realization of FIR as well as IIR analog adaptive filters were developed (Menzi, Zbinden and Moschytz, 1991; Menzi, 1992; Zbinden and Dąbrowski, 1992; Dąbrowski and Zbinden, 1993).

9.3.1 Adaptive FIR filters

A wide class of adaptive switched-capacitor filters approximating the optimum Wiener solution after the convergence of the adaptation process, follows from a variety of FIR SC filtering structures presented in Chapter 7. For adaptation, we are, however, restricted in practical SC realizations to the LMS algorithm as illustrated in Fig. 9.5a (Menzi, Zbinden and Moschytz, 1991; Menzi, 1992; Zbinden and Dąbrowski, 1992). To realize multiplications by the filter adaptive coefficients as well as multiplications in the LMS algorithm feedback loop, continuous-time four-quadrant multipliers (e.g. Hong, 1985; Song and Kim, 1990) as well as sigma-delta modulators (Huang and Moschytz, 1992) can be used.

9.3.2 Adaptive IIR filters

As already mentioned, the order of the adaptive system can be significantly decreased, resulting in a reduction of the chip area, power consumption, adaptation time, etc., if the conventional FIR configuration (Fig. 9.5a) is replaced by its IIR counterpart (Shynk, 1989). The best results can be obtained if the adaptive system is orthogonal, i.e. if the state signals (the signals that are multiplied by adaptive coefficients and then summed to form the output signal) are orthogonal (uncorrelated) if a white noise is put to the filter input (Johns, Snelgrove and Sedra, 1989, 1990). For an orthonormal system we assume additionally that all state signals represent the same average pseudopower (which can be normalized to unity). FIR filters are inherently orthonormal but only special IIR filter structures are orthonormal. An orthonormal IIR ladder structure that is suitable for SC realization is illustrated in Fig. 9.5b.

A very flexible solution was proposed by Zbinden and Dąbrowski (1992). The designed first-order adaptive filter module makes it possible to realize optimum, arbitrarily mixed FIR/IIR orthonormal structures. Delays, summers and integrators are realized as SC (i.e. discrete-time) circuits. On the other hand, multipliers are continuous-time circuits. All building-blocks are offset compensated. Clock-feedthrough of the integrator in the adaptive part is

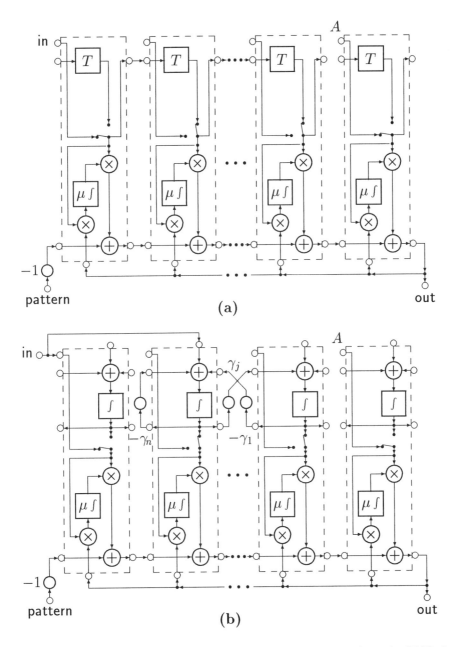

Fig. 9.5 Modular orthonormal adaptive filter structures based on the LMS algorithm with steplength μ: (a) FIR structure, (b) IIR structure; an additional module driven with arbitrary constant A compensates for filtering offset errors.

250 *Applications of multirate and multiphase SC circuits*

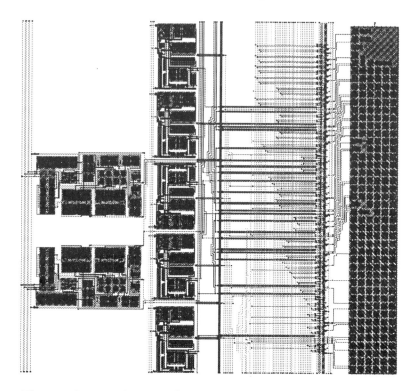

Fig. 9.6 Layout of the FIR/IIR adaptive filter module (SCADAP).

compensated for using an additional integrator with exactly the same structure but with an excitation equal to zero (Dąbrowski, 1993; Dąbrowski and Zbinden, 1993). The layout of this module is shown in Fig. 9.6.

References

Allen, P. and Sanchez-Sinencio, E. (1984) *Switched Capacitor Circuits*, Van Nostrand Reinhold, New York, NY.

Azizi, S.A. (1987) *Entwurf und Realisierung digitaler Filter*, Oldenbourg, Munich.

Balabanian, N. and Bickart, T.A. (1983) *Electrical Network Theory*, R. Krieger Publ. Co., Malabar, Florida.

Bedrosian, S.D. and Refai, S. (1982) Switched-capacitor networks: analysis and fault diagnosis. *Proc. 1982 IEEE Int. Large Scale Systems Symp.*, Virginia Beach, 316–20.

Bedrosian, S.D. and Refai, S. (1983) A flow graph transformation technique specifically for discrete time networks. *J. Franklin Inst.*, **315**(1), 27–36.

Belevitch, V. (1968) *Classical Network Theory*, Holden-Day, San Francisco, CA.

Bellanger, M.G. (1982) On computational complexity in digital transmultiplexer filters. *IEEE Trans. Comm.*, **COM-30**(7), 1461–5.

Bellanger, M.G. and Daguet, J.L. (1974) TDM-FDM transmultiplexer: digital polyphase and FFT. *IEEE Trans. Comm.*, **COM-22**(9), 1199–205.

Bellanger, M.G., Bonnerot, G. and Coudreuse, M. (1976) Digital filtering by polyphase network: application to sample rate alteration and filter banks. *IEEE Trans. Acoust. Speech Signal Process.*, **ASSP-24**(4), 109–14.

Berg, S.K., Hurst, P.J., Lewis, S.H. *et al.* (1994) A switched-capacitor filter in 2 μm CMOS using parallelism to sample at 80 MHz. *Digest Techn. Papers IEEE Int. Solid-State Circuits Conf.*, San Francisco, CA, 62–3.

Bon, M. and Kończykowska, A. (1981) All-symbolic analysis techniques for multiphase switched-capacitor networks. *Proc. European Conf. Circuit Theory Design*, The Hague, The Netherlands, 655–60.

Boser, B. and Wooley, B. (1988) The design of sigma-delta modulation analog-to-digital converters. *IEEE J. Solid State Circuits*, **SC-23**(6), 1298–308.

Brückmann, D. (1984) Design of VIS-SC filters simulating reference filters in lattice structure. *Arch. Elek. Übertrag.*, **38**(5), 327–30.

Brückmann, D. (1986) *Integrierte Schalter-Kondensator-Filter mit Spannungsumkehrschaltern*, Ruhr-Universität Bochum, Abteilung für Elektrotechnik, Bochum.

Bruton, L.T. and Vaughan-Pope, D.A. (1976) Synthesis of digital ladder filters from LC filters. *IEEE Trans. Circuits and Syst.*, **CAS-23**(6), 395–402.

Caves, J.T., Copeland, M.A., Rahim, C.F. and Rosenbaum, S.D. (1977) Sampled analog filtering using switched capacitors as resistor equivalents. *IEEE J. Solid State Circuits*, **SC-12**(6), 592–9.

Chui, C.K. (1992) *An Introduction to Wavelets*, Academic Press, San Diego, CA.

Ciota, Z. (1996) *Theory and Practical Realization of Analog Integrated Filters Particularly Taking into Account Finite Impulse Response Filters*, Łódź University of Technology Press, Łódź (in Polish).

Ciota, Z., Napieralski, A. and Noullet, J.L. (1993) Analog interpolated finite impulse response filters. *Proc. European Conf. Circuit Theory Design*, Davos, Switzerland, 1367–72.

Coates, C.L. (1959) Flow graph solutions of linear algebraic equations. *IRE Trans. Circuit Theory*, **CT-6**(6), 170–87.

Cohen, L. (1995) *Time-Frequency Analysis*, Prentice-Hall, Englewood Cliffs, NJ.

Cooley, J.W. and Tukey, J.W. (1965) An algorithm for the machine computation of complex Fourier series. *Math. Computation*, **19**(4), 227–301.

Coulson, A.J. (1995) A generalization of nonuniform bandpass sampling. *IEEE Trans. Signal Process.*, **43**(3), 694–704.

Crochiere, R.E. (1981) Subband coding. *Bell Syst. Tech. J.*, **60**(9), 1633–54.

Crochiere, R.E. and Rabiner, L.R. (1983) *Multirate Digital Signal Processing*, Prentice-Hall, Englewood Cliffs, NJ.

Crochiere, R.E., Webber, S.A. and Flanagan, J.L. (1976) Digital coding of speech in subbands. *Bell Syst. Tech. J.*, **55**(10), 1069–85.

Croisier, A., Esteban, D. and Galand, C. (1976) Perfect channel splitting by use of interpolation/decimation/tree decomposition techniques. *Proc. Symp. Info. Circuits and Syst.*, Patras, Greece, 191–5.

Dąbrowski, A. (1981) An efficient algorithm for solving state equations of nonlinear dynamic systems using heteromerous Runge-Kutta methods. *Proc. European Conf. Circuit Theory Design*, The Hague, The Netherlands, 329–34.

Dąbrowski, A. (1982a) Improved zero phase lag integrators with grounded switched capacitors. *Proc. IEEE Int. Symp. Circuits and Syst.*, Rome, Italy, 749–52.

Dąbrowski, A. (1982b) Synthesis of SC filters with grounded switched capacitors. *Proc. Summer Symp. Circuit Theory*, Prague, Czechoslovakia, 228–33.

Dąbrowski, A. (1982c) *Heteromerous Methods in Computer Aided Design of Electronic Circuits*, Poznań University of Technology, Poznań (in Polish).

Dąbrowski, A. (1983) Special techniques for efficient computer-aided transient analysis of electrical networks. *Proc. European Conf. Circuit Theory Design*, Stuttgart, Germany, 128–30.

Dąbrowski, A. (1985) Synthesis of switched-capacitor networks by two frequency transformations. *Proc. European Conf. Circuit Theory Design*, Prague, Czechoslovakia, 725–8.

Dąbrowski, A. (1986a) Design of multirate VIS-SC filters. *Proc. European Solid-State Circuits Conf.*, Delft, The Netherlands, 131–3.

Dąbrowski, A. (1986b) Suppression of parasitic oscillations in communication systems with wave digital filters. *Arch. Elek.*, **35**(3/4), 941–55.

Dąbrowski, A. (1987a) Transmultiplexers: technical requirements and realization methods. *Przegląd Telekomunikacyjny*, **LX**(7), 204–8 (in Polish).

Dąbrowski, A. (1987b) Wave digital filters and switched-capacitor filters for multistage transmultiplexers. *Proc. Nat. Symp. Telecom. KST'87*, Bydgoszcz, 135–44 (in Polish).

Dąbrowski, A. (1988a) Switched-capacitor interpolators and decimators using voltage inverter switches. *Arch. Elek. Übertrag.*, **42**(4), 217–26.

Dąbrowski, A. (1988b) *Recovery of Effective Pseudopower in Multirate Signal Processing*, Poznań University of Technology Press, Poznań (in Polish).

Dąbrowski, A. (1989a) Transmission of effective pseudopower in multirate signal processing. *Proc. IEEE Int. Conf. Acoust. Speech and Signal. Process.*, Glasgow, Scotland, UK, 1267–70.

Dąbrowski, A. (1989b) Principles of pseudo-N-path VIS-SC filters for multistage single-way transmultiplexers. *Proc. URSI Int. Symp. Signals, Systems and Electronics.*, Erlangen-Nürnberg, Germany, 69–72.

Dąbrowski, A. (1991) Losses and recovery of effective pseudopower in product modulation and multirate filtering. *Proc. Treizième Colloque GRETSI*, Juan-Les-Pins, France, 237–40.

Dąbrowski, A. (1993) Realization of adaptive IIR/FIR filters for communications using SC technique, in *Adaptive Methods and Emergent Techniques for Signal Processing and Communications* (eds D. Docampo and A.R. Figueiras), Universidad de Vigo, pp. 725–8.

Dąbrowski, A. (1995) *Application of Microwave Techniques in Digital Signal Processing*, Poznań University of Technology, TEMPUS JEP 4294 Press, Poznań (in Polish).

Dąbrowski, A. and Braszak, A. (1992) Adaptive infinite impulse response hybrids for primary rate ISDN. *Proc. European Signal Process. Conf.*, Brussels, Belgium, 127–30.

Dąbrowski, A. and Fettweis, A. (1986) Realization of multirate wave digital filters. *Proc. European Signal Process. Conf.*, The Hague, The Netherlands, 139–42.

Dąbrowski, A. and Fettweis, A. (1987) Generalized approach to sampling rate alteration in wave digital filters. *IEEE Trans. Circuits and Syst.*, **CAS-34**(6), 678–86.

Dąbrowski, A. and Moschytz, G.S. (1989) Direct by-inspection derivation of signal-flow graphs for multiphase stray-insensitive switched-capacitor networks. *Electron. Lett.*, **25**(6), 387–9.

Dąbrowski, A. and Moschytz, G.S. (1990a) Direct analysis of multiphase switched-capacitor networks using signal-flow graphs. *IEEE Trans. Circuits and Syst.*, **CAS-37**(5), 594–607.

Dąbrowski, A. and Moschytz, G.S. (1990b) Source-sink-node (SSN)-network transformation. *IEEE Trans. Circuits and Syst.*, **CAS-37**(5), 663–6.

Dąbrowski, A. and Zbinden, P. (1993) *SCADAP-Zell: Simulationen und Messungen*, Interner Bericht, ETH Zürich, Institut für Signal- und Informationsverarbeitung, Zürich.

Dąbrowski, A., Menzi, U. and Moschytz, G.S. (1989) Offset-compensated switched-capacitor delay circuit that is insensitive to stray capacitance and to capacitor mismatch. *Electron. Lett.*, **25**(6), 387–9.

Dąbrowski, A., Menzi, U. and Moschytz, G.S. (1992) Design of switched-capacitor FIR filters to a low-power MFSK receiver. *IEE Proc.*, part G, **139**(4), 450–66.

Deprettere, E.F. and Dewilde, P. (1980) Orthogonal cascade realization of real multiport digital filters. *Int. J. Circuit Theory Appl.*, **8**(3), 245–72.

Desoer, C.A. (1960) The optimum formula for the gain of a flow graph or a simple derivation of Coates' formula. *Proc. IRE*, **48**(5), 883–9.

Dias, V.F. and Franca, J.E. (1988) Parasitic-compensated switched-capacitor delay lines. *Electron. Lett.*, **24**(7), 377–9.

Dostál, T. (1994) On signal-flow graph analysis of multiphase SC networks. *Proc. IEEE Int. Symp. Circuits and Syst.*, London, Great Britain, 585–8.

Dostál, T. and Mikula, J. (1985) Analysis and synthesis of switched capacitor networks using flow graphs. *Proc. European Conf. Circuit Theory Design*, Prague, Czechoslovakia, 705–8.

Enomoto, T., Ishihara, T. and Yasumoto, M.A. (1982) Integrated tapped MOS analogue delay line using switched capacitor technique. *Electron. Lett.*, **18**(5), 193–4.

Esteban, D. and Galand, C. (1977) Application of quadrature mirror filters to split band voice coding schemes. *Proc. IEEE Int. Conf. Acoust. Speech and Signal. Process.*, Hartford, CT, 191–5.

Fettweis, A. (1960) Filters met willekeurig gekozen dempingspolen en Tschebyschewkarakteristiek in het doorlaatgebied. *Tijdschrift van het Nederlands radiogenootchap*, **25**(5–6), 337–82.

Fettweis, A. (1971) Digital filter structures related to classical filter networks. *Arch. Elek. Übertrag.*, **25**(2), 79–89.

Fettweis, A. (1972) Pseudopassivity, sensitivity and stability of wave digital filters. *IEEE Trans. Circuit Theory*, **CT-19**(6), 668–73.

Fettweis, A. (1973) Some general properties of signal-flow networks, in *Network and Signal Theory* (eds J.K. Skwirzyński and J.O. Scanlan), Peter Peregrinus, London, pp. 48–59.

Fettweis, A. (1974) On sensitivity and roundoff noise in wave digital filters. *IEEE Trans. Acoust. Speech Signal Process.*, **ASSP-22**(5), 383–84.

Fettweis, A. (1978a) Principles for multiplier-free single-way PCM/FDM and audio/FDM conversion. *Arch. Elek. Übertrag.*, **32**(11), 441–6.

Fettweis, A. (1978b) Multiplier-free modulation schemes for PCM/FDM and audio/FDM conversion. *Arch. Elek. Übertrag.*, **32**(12), 477–85.

Fettweis, A. (1979a) Basic principles of switched-capacitor filters using voltage inverter switches. *Arch. Elek. Übertrag.*, **33**(1), 13–19.

Fettweis, A. (1979b) Switched-capacitor filters using voltage inverter switches: further design principles. *Arch. Elek. Übertrag.*, **33**(3), 107–14.

Fettweis, A. (1982) Transmultiplexers with either analog conversion circuits, wave digital filters, or SC filters – a review. *IEEE Trans. Comm.*, **COM-30**(7), 1575–86.

Fettweis, A. (1984) Digital circuits and systems. *IEEE Trans. Circuits and Syst.*, **CAS-31**(1), 31–48.

Fettweis, A. (1986) Wave digital filters: theory and practice. *Proc. IEEE*, **74**(2), 270–327.

Fettweis, A. and Meerkötter, K. (1975) Suppression of parasitic oscillations in wave digital filters. *IEEE Trans. Circuits and Syst.*, **CAS-22**(3), 239–46.

Fettweis, A. and Meerkötter, K. (1977) On parasitic oscillations in digital filters under looped conditions. *IEEE Trans. Circuits and Syst.*, **CAS-24**(9), 475–81.

Fettweis, A. and Nossek, J.A. (1982) Sampling rate increase and decrease in wave digital filters. *IEEE Trans. Circuits and Syst.*, **CAS-29**(12), 797–806.

Fettweis, A. and Wupper, H. (1971) A solution to the balancing problem in N-path filters. *IEEE Trans. Circuit Theory*, **CT-18**(3), 403–5.

Fettweis, A., Herbst, D. and Nossek, J.A. (1979) Floating voltage inverter switches for switched-capacitor filters. *Arch. Elek. Übertrag.*, **33**(9), 376–7.

Fettweis, A., Herbst, D., Höfflinger, B. *et al.* (1980) MOS switched-capacitor filters using voltage inverter switches. *IEEE Trans. Circuits and Syst.*, **CAS-27**(6), 527–38.

Fettweis, A., Leickel, T., Bolle, M. and Sauvagerd, U. (1990) Realization of filter banks by means of wave digital filters. *Proc. IEEE Int. Symp. Circuits and Syst.*, New Orleans, 2013–6.

Fino, M.H., Franca, J.E. and Steiger-Garção, A. (1995a) Automatic symbolic characterization of SC multirate circuits with finite gain operational amplifiers. *Proc. IEEE Int. Symp. Circuits and Syst.*, Seattle, WA, 2213–16.

Fino, M.H., Franca, J.E. and Steiger-Garção, A. (1995b) Automatic symbolic analysis of switched-capacitor filtering networks using signal flow graphs. *IEEE Trans. Computer-Aided Design Int. Circuits and Syst.*, **14**(7), 858–67.

Fischer, G. (1985) *Design of SC Filters with Emphasis on High-Frequency Performance*, ETH Zürich, Institut für Signal- und Informationsverarbeitung, Zürich.

Fischer, G. (1987) Switched-capacitor FIR filters. *Proc. IEEE Int. Symp. Circuits and Syst.*, Philadelphia, PA, 742–5.

Fischer, G. (1990) Analog FIR filters by switched-capacitor techniques. *IEEE Trans. Circuits and Syst.*, **CAS-37**(6), 808–14.

Fleischer, P.E., Ganesan, A. and Laker, K.R. (1981) Parasitic compensated switched-capacitor circuits. *Electronics Lett.*, **17**(24), 929–31.

Fliege, N. (1973) Complementary transformation of feedback systems. *IEEE Trans. Circuits and Syst.*, **CAS-20**(3), 137–9.

Fliege, N. (1993) *Multiraten-Signalverarbeitung: Theorie und Anwendungen*, Teubner, Stuttgart.

Fliege, N. (1995) Fundamentals of multirate systems and filter banks, in *Microsystems Technology for Multimedia Applications: An Introduction* (eds B. Sheu *et al.*), IEEE Press, New York, NY, pp. 153–65.

Fraiture, L. and Neirynck, J.J. (1969) Theory of unit-element filters. *Revue H. F.*, **7**(7), 325–40.

Franca, J.E. (1984) Decimators and interpolators for narrowband switched-capacitor bandpass filter systems. *Proc. IEEE Int. Symp. Circuits and Syst.*, Montreal, Canada, 789–93.

Franca, J.E. (1985) Non-recursive polyphase switched-capacitor decimators and interpolators. *IEEE Trans. Circuits and Syst.*, **CAS-32**(9), 877–87.

Franca, J.E. and Haigh, D.G. (1988) Design and applications of single-path frequency translated switched-capacitor systems. *IEEE Trans. Circuits and Syst.*, **CAS-35**(4), 394–408.

Franca, J.E. and Martins, R.P. (1990) IIR switched-capacitor decimator building-blocks with optimum implementation. *IEEE Trans. Circuits and Syst.*, **CAS-37**(1), 81–90.

Franca, J.E. and Mitra, S.K. (1994) Design and application of analog–digital multirate signal processing, in *Circuits and Systems – Tutorials* (eds C. Toumazou, N. Battersby and S. Porta), IEEE ISCAS'94, London, pp. 71–106.

Franca, J.E. and Santos, S. (1988) FIR switched-capacitor decimators with active-delayed block polyphase structures. *IEEE Trans. Circuits and Syst.*, **CAS-35**(8), 1033–7.

Franks, L.E. and Sandberg, I.W. (1960) An alternative approach to the realization of network transfer functions: the N-path filter. *Bell Syst. Tech. J.*, **39**(5), 1321–50.

Gazsi, L. (1984) *Reference Manual for FALCON Program*, Ruhr-Universität Bochum, Lehrstuhl für Nachrichtentechnik, Bochum.

Gazsi, L. (1985) Explicit formulas for lattice wave digital filters. *IEEE Trans. Circuits and Syst.*, **CAS-32**(1), 68–88.

Gear, C. W. (1971) *Numerical Initial Value Problems in Ordinary Differential Equations*, Prentice-Hall, Englewood Cliffs, NJ.

Gersho, A. (1978) Principles of quantization. *IEEE Trans. Circuits and Syst.*, **CAS-25**(7), 427–36.

Ghaderi, M.B., Temes, G.C. and Law, S. (1981) Linear interpolation using CCD's or switched-capacitor filters. *IEE Proc.*, part G, **128**(4), 213–5.

Gillingham, P. (1984) Stray-free switched-capacitor unit-delay circuit. *Electron. Lett.*, **20**(7), 308–10.

Givens, W. (1958) Computation of plane unitary rotations transforming a general matrix to triangular form. *J. Soc. Indust. Appl. Math.*, **6**(1), 26–50.

Gray, A.H. (1980) Passive cascaded lattice digital filters. *IEEE Trans. Circuits and Syst.*, **CAS-27**(5), 337–44.

Gray, A.H. and Markel, J.D. (1973) Digital lattice and ladder filter synthesis. *IEEE Trans. Audio Electroacoust.*, **AU-21**(6), 491–500.

Gregorian, R. and Nicholson, W.E. (1980) Switched-capacitor decimation and interpolation circuits. *IEEE Trans. Circuits and Syst.*, **CAS-27**(6), 509–14.

Gregorian, R., Martin, K.W. and Temes, G.C. (1983) Switched-capacitor circuit design. *Proc. IEEE*, **71**(8), 941–66.

Grimbleby, J.B. (1990) *Computer-aided Analysis and Design of Electronic Networks*, Pitman, London.

Grünigen, D.C. (1983) *Entwurfs- und Analyseverfahren für Filter mit geschalteten Kondensatoren*, ETH Zürich, Institut für Signal- und Informationsverarbeitung, Zürich.

Grünigen, D.C., Brugger, U.W. and Moschytz, G.S. (1981) A simple switched-capacitor decimation circuit. *Electron. Lett.*, **17**(1), 30–1.

Grünigen, D.C., Brugger, U.W., Vollenweider, W. *et al.* (1981) Combined switched-capacitor FIR N-path filter using only grounded capacitors. *Electron. Lett.*, **17**(1), 788–90.

Grünigen, D.C., Sigg, R., Ludwig, M. *et al.* (1982) Integrated switched-capacitor low-pass filter with combined anti-aliasing decimation filter for low frequencies. *IEEE J. Solid State Circuits*, **SC-17**(6), 1024–9.

Haigh, D.G., Toumazou, C., Harrold, S.J. *et al.* (1991) Design optimization and testing of a GaAs switched-capacitor filter. *IEEE Trans. Circuits and Syst.*, **38**(8), 825–37.

Handkiewicz, A. (1988) *Effective Synthesis Algorithms for Switched-Capacitor Filters*, Poznań University of Technology Press, Poznań (in Polish).

Harrold, S.J., Vance, I.A.W. and Haigh, D.G. (1985) Second-order switched-capacitor bandpass filter implemented in GaAs. *Electron. Lett.*, **21**(11), 494–6.

Hasler, M. (1981) Stray capacitance insensitive switched capacitor filters. *Proc. IEEE Int. Symp. Circuits and Syst.*, Chicago, IL, 42–5.

Herbst, D., Höfflinger, B., Schumacher, K. et al. (1979) MOS switched-capacitor filters with reduced number of operational amplifiers. *IEEE J. Solid State Circuits*, **SC-14**(6), 1010–19.

Herbst, D., Pandel, J., Fettweis, A. et al. (1982) VIS-SC-filters with reduced influences of parasitic capacitances. *IEE Proc.*, part G, **129**(2), 29–39.

Hökenek, E. and Moschytz, G.S. (1980) Analysis of multiphase switched-capacitor (m.s.c.) networks using the indefinite admittance matrix (i.a.m.). *IEE Proc.*, part G, **127**(5), 226–41.

Hökenek, E. and Moschytz, G.S. (1983) Design of parasitic-insensitive bilinear-transformed admittance-scaled (BITAS) SC ladder filters. *IEEE Trans. Circuits and Syst.*, **CAS-30**(12), 873–88.

Hong, Z. (1985) *Analogue Four-Quadrant Multiplier in CMOS Technology*, ETH Zürich, Abteilung für Elektrotechnik, Zürich.

Hostička, B.J., Brodersen, R.W. and Gray, P.R. (1977) MOS sampled data recursive filters using switched-capacitor integrators. *IEEE J. Solid State Circuits*, **SC-12**(6), 600–8.

Huang, Q. and Moschytz, G.S. (1992) Multiplierless analog LMS adaptive FIR filters. *Proc. IEEE Int. Symp. Circuits and Syst.*, San Diego, CA, 749–52.

Hurst, P.J. (1991) Shifting the frequency response of switched-capacitor filters by nonuniform sampling. *IEEE Trans. Circuits and Syst.*, **CAS-38**(1), 12–19.

Jain, V.K. and Crochiere, R.E. (1984) Quadrature mirror filter design in the time domain. *IEEE Trans. Acoust. Speech Signal Process.*, **ASSP-32** (4), 353–61.

Jayant, N.S. and Noll, P. (1984) *Digital Coding of Waveforms*, Prentice-Hall, Englewood Cliffs, NJ.

Jerri, A.J. (1977) The Shannon sampling theorem – its various extensions and applications: a tutorial review. *Proc. IEEE*, **65**(11), 1565–96.

Johns, D.A., Snelgrove, M. and Sedra, A.S. (1989) Orthonormal ladder filters. *IEEE Trans. Circuits and Syst.*, **CAS-36**(3), 337–43.

Johns, D.A., Snelgrove, M. and Sedra, A.S. (1990) Adaptive recursive state-space filters using gradient-based algorithm. *IEEE Trans. Circuits and Syst.*, **CAS-37**(6), 673–84.

Johnston, J.D. (1980) A filter family designed for use in quadrature mirror filter banks. *Proc. IEEE Int. Conf. Acoust. Speech and Signal. Process.*, Denver, CO, 291–4.

Kaelin, A. (1988) Exakter systematischer Entwurf von streuinsensitiven, modularen, elliptischen SC-Kettenfiltern. *Mitteilungen AGEN*, **47**(5), 9–16.

Kleine, U. (1985) *Integrierte Schalter-Kondensator-Filter in CMOS Technologie*, Universität Dortmund, Abteilung für Elektrotechnik, Dortmund.

Kleine, U., Herbst, D., Höfflinger, B. *et al.* (1981) Real-time programmable unit-element SC filter for LPC synthesis. *Electron. Lett.*, **17**(17), 600–2.

Kohlenberg, A. (1953) Exact interpolation of band-limited functions. *J. Appl. Physics*, **24**(12), 1432–6.

Kończykowska, A. and Bon, M. (1988) Automated design software for switched-capacitor IC's with symbolic simulator SCYMBAL. *Proc. 25th ACM/IEEE Design Automation Conf.*, Anaheim, CA, 363–8.

Korzec, Z. and Ciota, Z. (1991) FIR filter realization by multiphase SC techniques. *Proc. XIVth National Conf. Circuit Theory and Electronic Systems*, Waplewo, Poland, 345–50.

Korzec, Z. and Tounsi, L. (1994) Analysis of nonuniformly sampled switched-capacitor filters. *Proc. Int. Conf. VLSI*, Dębe, Poland, 124–9.

Kotelnikov, V.A. (1933) On the transmission capacity of 'ether' and wire in electrocommunications. *Proc. First all-Union Conf. Questions of Comm.*, Moscow, 63–8.

Lee, Y.S. and Martin, K.W. (1988) A switched-capacitor realization of multiple FIR filters on a single chip. *IEEE J. Solid State Circuits*, **SC-23**(4), 536–42.

Lee, M.S., Temes, G.C., Chang, C. and Ghaderi, M.B. (1981) Bilinear switched-capacitor ladder filters. *IEEE Trans. Circuits and Syst.*, **CAS-28**(8), 811–22.

Leontiev, A.P. (1976) *Exponential series*, Fizmat, Moscow (in Russian).

Lin, P.M. (1991) *Symbolic Network Analysis*, Elsevier, Amsterdam.

Mallat, S.G. (1989) Theory of multiresolution signal decomposition: the wavelet representation. *IEEE Trans. Pattern Anal. Machine Intell.*, **11**(7), 674–93.

Maloberti, F. (1991) Over sampling converters, in *Analogue–Digital ASICs* (eds R. S. Soin, F. Maloberti and J.E. Franca), Peter Peregrinus, London, pp. 143–72.

Martin, K. (1980) Improved circuits for the realization of switched-capacitor filters. *IEEE Trans. Circuits and Syst.*, **CAS-27**(4), 237–44.

Martins, R.P. and Franca, J.E. (1989) a 2.4 μm CMOS switched-capacitor video decimator with sampling rate reduction from 40.5 MHz to 13.5 MHz. *Proc. IEEE Custom Integrated Circuits Conf.*, San Diego, CA, 25.4.1–4.

Martins, R.P. and Franca, J.E. (1991) An experimental 1.8 μm CMOS anti-aliasing switched-capacitor decimator with high input sampling frequency. *Proc. European Solid-State Circuits Conf.*, Milan, Italy, 13–16.

Martins, R.P. and Franca, J.E. (1995) Cascade switched-capacitor IIR decimating filters. *IEEE Trans. Circuits and Syst.–1: Fund. Theory Appl.*, **42**(7), 367–76.

Mason, S.J. (1953) Feedback theory – some properties of signal flow graphs. *Proc. IRE*, **41**(9), 1144–56.

Mason, S.J. (1956) Feedback theory – further properties of signal flow graphs. *Proc. IRE*, **44**(7), 920–6.

Matsui, K., Matsuura, T., Fukasawa, S. et al. (1985) CMOS video filters using switched-capacitor circuits. *IEEE J. Solid State Circuits*, **SC-20**(12), 1096–102.

Mavor, J., Reekie, H.M., Denyer, P.B. et al. (1981) A prototype switched-capacitor voltage-wave filter realized in NMOS technology. *IEEE J. Solid State Circuits*, **SC-16**(6), 716–23.

McClellan, J.H. and Parks, T.W. (1973) A unified approach to the design of optimum FIR linear-phase digital filters. *IEEE Trans. Circuit Theory*, **CT-20**(6), 697–701.

McConnell, E. (1995) Advanced data acquisition sampling techniques: seamless changing of the sample rate and equivalent time sampling. *Proc. Data Acquisition Conf.*, Boston, MA, 1–6.

McWhirter, J.G. and Proudler, I.K. (1992) Orthogonal lattice algorithms for adaptive filtering and beamforming, in *Algorithms and Parallel VLSI Architectures* (eds P. Quinton and Y. Robert), Elsevier, Amsterdam, pp. 11–23.

Menzi, U. (1992) *Switched-Capacitor Realisierung von Estimations-Algorithmen für Adaptive Filter und MFSK Empfänger*, ETH Zürich, Institut für Signal- und Informationsverarbeitung, Zürich.

Menzi, U., Zbinden, P. and Moschytz, G.S. (1991) Adaptive switched-capacitor filters based on the LMS algorithm. *Proc. European Conf. Circuit Theory Design*, Copenhagen, Denmark, 1211–20.

Montecchi, F. and Maloberti, F. (1983) Switched-capacitor ladder filters for high frequency applications via bilinear resistor modeling. *Proc. 26th Midwest Symp. Circuits and Syst.*, Mexico, 445–8.

Moschytz, G.S. (1974) *Linear Integrated Networks: Fundamentals*, Van Nostrand Reinhold, New York, NY.

Moschytz, G.S. (1976) The morphological approach to network and circuit design. *IEEE Trans. Circuits and Syst.*, **CAS-23**(4), 239–42.

Moschytz, G.S. (1987) Elements of four-port matrix theory as required for SC network analysis. *Int. J. Circuit Theory Appl.*, **15**(3), 235–49.

Moschytz, G.S. and Brugger, U.W. (1984) Signal-flow graph analysis of SC networks. *IEE Proc.*, part G, **131**(2), 72–85.

Moschytz, G.S. and Mulawka, J.J. (1986) Direct analysis of stray-insensitive switched-capacitor networks using signal flow graphs. *IEE Proc.*, part G, **133**(3), 145–53.

Mulawka, J.J. (1981) Switched-capacitor analogue delays comprising unity gain buffers. *Electron. Lett.*, **17**(7), 276–7.

Mulawka, J.J. and Moschytz, G.S. (1985) Direct analysis of stray-insensitive switched-capacitor networks using signal flow graphs. *IEE Proc.*, part G, **132**(6), 255–65.

Nagaraj, K. (1984) Switched-capacitor delay circuit that is insensitive to capacitor mismatch and stray capacitance. *Electron. Lett.*, **20**(16), 663–4.

Napieralski, A., Noullet, J.L. and Ciota, Z. (1994) Realization of some different FIR SC filters in the CMOS technology. *Bull. Polish Acad. Sci. – Tech. Sci.*, **42**(2), 269–78.

Nathan, A. (1961) Matrix analysis of networks having infinite-gain operational amplifiers. *Proc. IRE*, **49**(10), 1577–8.

Neuvo, Y., Dong, C.-Y. and Mitra, S.K. (1984) Interpolated finite impulse response filters. *IEEE Trans. Acoust. Speech Signal Process.*, **ASSP-32**(3), 563–70.

Nossek, J.A. and Temes, G.C. (1980) Switched-capacitor filter design using bilinear element modeling. *IEEE Trans. Circuits and Syst.*, **CAS-27**(6), 481–91.

Nyquist, H. (1928) Certain topics in telegraph transmission theory. *Trans. AIEE*, **47**(6), 617–64.

Oppenheim, A.V. and Schafer, R.W. (1975) *Digital Signal Processing*, Prentice-Hall, Englewood Cliffs, NJ.

Orchard, H.J. (1966) Inductorless filters. *Electron. Lett.*, **2**(6), 224–5.

Ozaki, H. and Ishii, J. (1958) Synthesis of a class of strip-line filters. *IRE Trans. Circuit Theory*, **CT-5**(6), 104–9.

Pain, B.G. (1979) Alternative approach to the design of switched-capacitor filters. *Electron. Lett.*, **15**(14), 438–9.

Pandel, J. (1981) Switched-capacitor elements for VIS-SC-filters with reduced influences of parasitic capacitances. *Arch. Elek. Übertrag.*, **35**(3), 121–30.

Pandel, J. (1982) Principles of pseudo-N-path switched-capacitor filters using recharging devices. *Arch. Elek. Übertrag.*, **36**(5), 177–87.

Pandel, J. (1983a) Design of higher order pseudo-N-path filters and frequency symmetrical filters in VIS-SC technique. *Arch. Elek. Übertrag.*, **37**(7/8), 251–60.

Pandel, J. (1983b) *Entwurf von Schalter-Kondensator-Filter mit Spannungsumkehrschaltern*, Ruhr-Universität Bochum, Abteilung für Elektrotechnik, Bochum.

Pandel, J. and Dawei, H. (1981) Novel realization of floating voltage inverter switches and inductances for switched-capacitor filters. *Proc. 15th Asilomar Conf. Circuits, Syst. and Computers*, Pacific Grove, CA, USA, 293–7.

Pandel, J., Brückmann, D., Fettweis, A. *et al.* (1986) Integrated 18th order pseudo-N-path filter in VIS-SC technique. *IEEE Trans. Circuits and Syst.*, **CAS-33**(2), 158–66.

Petraglia, A. and Mitra, S.K. (1992) High speed A/D conversion incorporating a QMF bank. *IEEE Trans. Instrumentation and Measurement*, **IM-41**(6), 427–31.

Petraglia, A. and Mitra, S.K. (1993) Design of magnitude preserving analog-to-digital converter. *IEICE Trans. Fundamentals*, **E76-A**(2), 149–55.

Piekarski, M.S. and Zarzycki, J. (1986) Orthogonal digital filtering. *Proc. IXth National Conf. Circuit Theory and Electron. Syst.*, Wrocław, 5–15 (in Polish).

Rabiner, L.R. and Gold, B. (1975) *Theory and Application of Digital Signal Processing*, Prentice-Hall, Englewood Cliffs, NJ.

Ragazzini, J.R. and Franklin, G.F. (1958) *Sampled-Data Control Systems*, McGraw-Hill, New York, NY.

Richards, P.L. (1948) Resistor-transmission-line circuits. *Proc. IRE*, **36**(2), 217–20.

Roy, R. and Lowenschuss, O. (1970) Design of MTI filters with non-uniform interpulse periods. *IEEE Trans. Circuit Theory*, **CT-17**(6), 604–12.

Saal, R. (1979) *Handbook of Filter Design*, AEG Telefunken, Berlin.

Said, A.E. (1985) Stray-free switched-capacitor building-block that realizes delay, constant multiplier, or summer circuit. *Electron. Lett.*, **21**(4), 167–8.

Saramäki, T., Neuvo, Y. and Mitra, S.K. (1988) Design of computationally efficient interpolated FIR filters. *IEEE Trans. Circuits and Syst.*, **CAS-35**(1), 70–88.

Shannon, C.E. (1949) Communication in the presence of noise. *Proc. IRE*, **37**(1), 10–21.

Shynk, J.J. (1989) Adaptive IIR filtering. *IEEE Acoust. Speech Signal Process. Magazine*, **6**(2), 4–21.

Smith, M.J.T. and Barnwell III, T.P. (1984) A procedure for designing exact reconstruction filter banks for structured subband coders. *Proc. IEEE Int. Conf. Acoust. Speech and Signal Process.*, San Diego, CA, 27.1.1–4.

Soin, R.S., Morris, S. and Maloberti, F. (1991) Some applications of mixed signal ASICs, in *Analogue–Digital ASICs* (eds R. S. Soin, F. Maloberti and J.E. Franca), Peter Peregrinus, London, pp. 239–72.

Somayazulu, V.S., Mitra, S.K. and Shynk, J.J. (1989) Adaptive line enhancement using multirate techniques. *Proc. IEEE Int. Conf. Acoust. Speech and Signal. Process.*, Portland, OR, 928–31.

Song, H. and Kim, C. (1990) An MOS four-quadrant analog multiplier using two-input squaring circuits with source followers. *IEEE J. Solid State Circuits*, **SC-25**(6), 1298–308.

Suyama, K. and Fang, S.C. (1992) *User's Manual for SWITCAP2*, Columbia University Press, New York, NY.

Taylor, R.C and Horvat, H. (1993) A high precision multi-notch low-pass switched-capacitor filter. *Proc. Fourth Mid-European Conference Custom Application Specific Integrated Circuits*, Budapest, 53–60.

Temes, G.C., Orchard, H.J and Jahanbegloo, M. (1978) Switched-capacitor filter design using the bilinear $\mathcal{Z}$-transform. *IEEE Trans. Circuits and Syst.*, **CAS-25**(12), 1039–44.

Toumazou, C. and Haigh, D.G. (1987) Design of a high-gain, single-stage operational amplifier for GaAs switched-capacitor filters. *Electron. Lett.*, **23**(14), 752–4.

Treichler, J.R. (1987) *Theory and Design of Adaptive Filters*, John Wiley, New York, NY.

Tsividis, Y.P. (1983) Representation of sampled-signals as functions of continuous time. *Proc. IEEE*, **71**(1), 181–3.

Uehara, C.T. and Gray, P.R. (1994) A 100 MHz output rate analog-to-digital interface for PRML magnetic-disk read channels in 1.2μ m CMOS. *Digest Techn. Papers IEEE Int. Solid-State Circuits Conf.*, San Francisco, CA, 280–1.

Unbehauen, R. and Cichocki, A. (1989) *MOS Switched-Capacitor and Continuous-Time Integrated Circuits and Systems*, Springer-Verlag, Berlin.

Vaidyanathan, P.P. (1985) The discrete-time bounded-real lemma in digital filtering. *IEEE Trans. Circuits and Syst.*, **CAS-32**(9), 918–24.

Vaidyanathan, P.P. (1986) Passive cascaded lattice structures for low sensitivity FIR filter design, with applications to filter banks. *IEEE Trans. Circuits and Syst.*, **CAS-33**(11), 1045–64.

Vaidyanathan, P.P. (1993) *Multirate Systems and Filter Banks*, Prentice-Hall, Englewood Cliffs, NJ.

Vaidyanathan, P.P. (1995) Sampling theorems from wavelet and filter bank theory, in *Microsystems Technology for Multimedia Applications: An Introduction* (eds B. Sheu *et al.*), IEEE Press, New York, NY, pp. 179–89.

Vaidyanathan, P.P. and Mitra, S.K. (1985) Passivity properties of low-sensitivity digital filter structures. *IEEE Trans. Circuits and Syst.*, **CAS-32**(3), 217–24.

Vaidyanathan, P.P. and Phoong, S-M. (1995) Reconstruction of sequences from nonuniform samples. *Proc. IEEE Int. Symp. Circuits and Syst.*, Seattle, 601–4.

Vandewalle, J., De Man, H.J. and Rabaey, J. (1981) Time, frequency, and z-domain modified nodal analysis of switched-capacitor networks. *IEEE Trans. Circuits and Syst.*, **CAS-28**(3), 186–95.

Vaughan, R.G., Scott, N.L. and White, D.R. (1991) The theory of bandpass sampling, *IEEE Trans. Signal Process.*, **39**(9), 1973–84.

Vetterli, M. (1986a) Filter banks allowing for perfect reconstruction. *Signal Process.*, **10**(4), 219–44.

Vetterli, M. (1986b) Perfect transmultiplexers. *Proc. IEEE Int. Conf. Acoust. Speech and Signal Process.*, Tokyo, 2567–70.

Vetterli, M. (1987) A theory of multirate filter banks. *IEEE Trans. Acoust. Speech Signal Process.*, **ASSP-35**(3), 356–72.

Vetterli, M. and Herley, C. (1992) Wavelets and filter banks: theory and design. *IEEE Trans. Signal Process.*, **SP-40**(9), 2207–32.

Vittoz, E.A. (1990) Future of analog in the VLSI environment. *Proc. IEEE Int. Symp. Circuits and Syst.*, New Orleans, 1372–5.

Vlach, J., Vlach, M. and Singhal, K. (1984) *WATSCAD Tutorial* and *WATSCAD User's Manual*, University of Waterloo, Ontario.

Walter, G.C. (1992) A sampling theorem for wavelet subspaces. *IEEE Trans. Inf. Theory*, **38**(2), 881–4.

Whittaker, E.T. (1915) On the functions which are represented by the expansion of interpolating theory. *Proc. Roy. Soc. Edinburgh*, **35**(2), 181–94.

Whittaker, J.M. (1929) The Fourier theory of the cardinal functions. *Proc. Math. Soc. Edinburgh*, **1**(2), 169–76.

Wojtkiewicz, A. and Tuszyński, M. (1983) $\mathcal{Z}$-domain matrix analysis of nonuniformly sampled data processing by digital filters. *Proc. European Conf. Circuit Theory Design*, Stuttgart, Germany, 423–5.

Wojtkiewicz, A., Tuszyński, M. and Klimkiewicz, W. (1985) Analysis and design of MTI digital filters processing nonuniformly sampled signals. *Proc. European Conf. Circuit Theory Design*, Prague, Czechoslovakia, 667–72.

Zadeh, L. (1950) Frequency analysis of variable networks. *Proc. IRE*, **38**(3), 291–9.

Zayed, A.I. (1993) *Advances in Shannon's Sampling Theory*, CRC Press, Boca Raton, Florida.

Zbinden, P. (1993) *Computergestützte Erzeugung und Optimierung von arrayartigen Layouts für allgemeine SC-Netzwerke und -Filter*, Hartung-Gorre, Konstanz.

Zbinden, P. and Dąbrowski, A. (1992) *Testintegration: Adaptives Modul, Hochpassfilter, Filter mit verschiedenen Verstärkern*, Interner Bericht, ETH Zürich, Institut für Signal- und Informationsverarbeitung, Zürich.

Zhou, X-F. and Shen, Z-G. (1985) Signal flow graph analysis of switched-capacitor networks. *Int. J. Circuit Theory Appl.*, **13**(4), 179–89.

Zwicky, F. (1967) The morphological approach to discovery, invention, research, and construction, in *New Methods of Thought and Procedure* (eds F. Zwicky and A.G. Wilson), Springer-Verlag, Berlin.

Index

AAF, *see* Anti-aliasing filter
Active circuits
   low-voltage 4
   low-power 4, 247
   see also Integrated circuits
Active-delayed block (ADB) structure 197–8, 200
A/D, *see* Analog-to-digital converter
Adaptive filter 169, 182, 241, 247–60
   orthonormal FIR/IIR structure 248, **249**
Adaptor 55
   see also Wave digital filter
ADB structure, *see* Active-delayed block structure
ADC, *see* Analog-to-digital converter
Adder, *see* Summer circuit
AIF, *see* Anti-imaging filter
Aliasing effect, *see* Nonrecoverable aliasing
Allen, P. 182, 251
All-pass transfer function 97, 104–5, 113–7, 203
Alteration of sampling rate, *see* Sampling, rate alteration
Amplifiers 27–30
   see also Operational amplifier, Operational transconductance amplifier and Buffer

Analog interface 6
Analog-to-digital (A/D) converter 34, 94, 244, 246
Analysis filter bank 14, 60, **91**, 92, **94**, 96–115, 246, **247**
Analysis of SC networks
   circuit analysis 129–33
   signal-flow graph approach 134–68
   by-inspection 134
Analytic function 45
Angular measure
   absolute 35n, 44
   relative 35n, 44
Anti-aliasing filter (AAF) 5, 36, 65, 243–4, 245
Anti-imaging filter (AIF) 5, 36, 62, 64, 243–4
Application specific integrated circuit (ASIC) 5, 244
ASIC, *see* Application specific integrated circuit
Attenuation, *see* Filter, attenuation
Azizi, S.A. 38, 251

Backward Euler transformation *206*, 210, 211
Balabanian, N. 1, 251
Bank, filter 5, 14, 58, 91–111
   analysis, *see* Analysis filter bank
   branching, *see* Branching fil-

272   *Index*

ter bank
  nonuniform (octave) 91, **92**
  polyphase, *see* Polyphase, filter bank
  power complementary, *see* Quadrature-mirror filter bank
  QMF, *see* Quadrature-mirror filter bank
  synthesis, *see* Synthesis filter bank
  two-channel 61, 93
  uniform 91, **92**
Barnwell III, T.P. 106, 267
Bedrosian–Refai formula 142, 143
Bedrosian, S.D. 129, 134, 135, 138, 167, 251
Belevitch, V. 59, 75, 117, 118, 251
Bellanger, M.G. 93, 109, 251
Berg, S.K. 14, 251
Bickart, T.A. 1, 251
BiCMOS technology, *see* Bipolar/complementary metal-oxide-semiconductor technology
Bilinear transformation 50, 204, 205, 207
Bipolar/complementary metal-oxide-semiconductor (BiCMOS) technology 6
Bireciprocal scheme 114, 116
Bolle, M. 91, 257
Bon, M. 129, 134, 252, 262
Bonnerot, G. 109, 251

Boser, B. 246, 252
Branching filter bank 59, 60, 93–5
Braszak, A. 247, 255
Brodersen, R.W. 5, 261
Brückmann, D. 112, 215, 217, 218, **219**, **220**, 252, 266
Brugger, U.W. 14, 129, 134, 260, 265
Bruton, L.T. 55, 252
Buffer 2, **162**, 173

Canonic
  impedance, *see* Impedance, canonic
  reactance, *see* Reactance, canonic
  reflectance, *see* Reflectance, canonic
Capacitance
  fringe 23, 25, 26
  matrix 136
  parasitic 15, 22, 23, 26, 186
    bottom plate 15, 22, 26
    op amp 15
    switch 15
    top plate 15, 22, 26
  (precise) ratio 5, 10, 25, 120, 127
  unit 23, 24
  see also Capacitor
Capacitor
  bank 25, 26
  bottom plate 22
  dummy 25, 26
  feedback (accumulator) 28n,

Index 273

174, 201
layout 23, **24**
matched 24
microphotograph **26**
miniature 3
multiphase switched 147, 149, 164
non-unit 23, 24, 26
  see also Stub
storage 174
top plate 22
unit 23, 24, 26
see also Capacitance
Carrier
  sequence, see Sequence, modulation
  signal, see Signal, carrier
Cauer filter 235, **237**, 240
Causality 51
Caves, J.T. 5, 252
CCD, see Charge-coupled device
Chang, C. 204, 263
Characteristic constant 50
Characteristic function 116–7
Charge
  injection 20
  vector 136
  preservation principle 130, 136
Charge-coupled device (CCD) 7
Charge-transfer device (CTD) 7, 8
Chui, C.K. 13, 40, 111, 252
Cichocki, A. 129, 268

Ciota, Z. 26, 169, 202, 252, 262, 265
Circuits
  continuous-time 4, 244, 245, 248
  multiphase 13, 14
  multirate 13, 247
  sampled-data 4
  switched-capacitor (SC) 5–12
  see also Integrated circuits
Circular shift 72
Classical sampling theorem 39, 71
Clock 19–21
  complementary signals 16
  feedthrough 17, 20, 248
  four-phase **19**, 133
  heteromerous 200
  high-frequency 21
  logical **19–20**, 21
  multiphase 14, **20**
    symmetrical (uniform) 21, 135
  pulse-width 28
  two-phase 6, **19**, 200, 206
CMOS technology, see Complementary metal-oxide-semiconductor technology
Coates, C.L. 135, 252
Coates–Desoer formula 142, 143
Cohen, L. 13, 252
Commensurate sampling, see Sampling, commensurate
Communication link 55, 224
Complementary filters 59, 98,

274 *Index*

107
half-band (symmetrical) 99–100
valid half-band 100, 107
see also Branching filter bank, Quadrature-mirror filter bank and Power complementary condition
Complementary metal-oxide-semiconductor (CMOS) technology 6, 26, 27, 170, 196
Complementary transformation 157
Complex frequency 50
Complex rotation 55, 58
Compression
  data, *see* Data, compression
  sampling rate, *see* Down-sampling
Continuous-time
  signal 31, 36, 38–42, 46, 68
  system 50, 52, 245
Controlled source 2
Conventional $\mathcal{Z}$ transfer function, *see* Transfer function, conventional $\mathcal{Z}$
Conversion function 77–81, 229–31
  aliasing 96–7
  cut 80
  fitted 78, 80
  main 77
  modified 80
  normalized 229–30
  resultant 81, 230–1

Cooley, J.W. 55, 252
Copeland, M.A. 252
Coudreuse, M. 109, 251
Coulson, A.J. 40, 253
Crochiere, R.E. 12, 40, 91, 101, 253, 261
Croisier, A. 99, 253
CTD, *see* Charge-transfer device

D/A, *see* Digital-to-analog converter
Dąbrowski, A. 6, 13n, 14, 29, 32, 40, 51, 55, 93, 112, 117, 118, 119, 129, 131, 134, 153, 155, 169, 170, 177, 178, 182, 187, 196, 204, 205, 211, 217, 221, 224, 228, 231, 241, 243, 247, 248, 250, 253–5, 270
Daguet, J.L. 93, 251
Data
  acquisition 119
  compression 13
  converters 241, 244–5
  oversampling 245–6
  polyphase 246
  see also Analog-to-digital and Digital-to-analog converters
  transmission 13
Dawei, H. 215, 217, 266
DC offset 20
Decimation 12, 62, 65–7, 80, 224, 227–30
  cascaded structure 203

see also Decimator
Decimator 14, 20, 65, 104, 170, 203, 221, 225, 244
  FIR SC 199–200
  heteromerous **123**
  polyphase structure **89**, 90, 109
  scheme **64**
  see also Decimation
Decoder 94
Delay element 15, 170, 174, 177–82, 248
  even–odd 132–3, 153, 177–9
  Gillingham 177, **180**, 181–2, 187, **189**
    generalized **183**
    modified four-phase **197**–8
  Nagaraj 130–2, 177
Delay factor 43
  see also $\mathcal{Z}$ transform
Delay-free loop 49n, 50, 51
Delay line **171**, 174, **179**, 180, 181, **188**–**91**
  see also Finite impulse response filter
De Man, H.J. 129, 269
Denyer, P.B. 51, 204, 226, 264
Deprettere, E.F. 55, 111, 255
Desoer, C.A. 135, 255
Dewilde, P. 55, 111, 255
DFT, *see* Discrete Fourier transformation
Dias, V.F. 177, 256
Digital
  filter, *see* Filter, digital
  signal processor (DSP) 5, 8, 9
Digital-to-analog (D/A) converter 244, 246
Dirac pulse 39, 40, 42, 71
Directional filter bank, *see* Branching filter bank
Dirichlet transform 46, 121, 122, 126
Dirichlet transformation 45–6
Discrete Fourier transformation (DFT) 37–8, 53, 57, 58, 70, 104, 105, 110
  two-point **103**, 104, **113**, **115**
Discrete-time
  current 51
  reference circuit 50, 51–2
  signal 31–48, 69
  system 31–60
  voltage 51
Distortion transfer function 96–106
Dong, C.-Y. 201, 265
Doppler signal 119
Dostál, T. 134, 256
Double-frequency transformation 205–11
Down-sampler 65, 82, 90, 95, 96
  simple equivalences **83**
Down-sampling 65, 73
DSP, *see* Digital signal processor
$\mathcal{D}$ transform, *see* Dirichlet transform
$\mathcal{D}$ transformation, *see* Dirichlet

transformation
Dynamic system 51, 119

Effective
  current 212
  delay 179, 201n
  pseudoenergy, *see* Pseudo-energy, effective
  resistance 212
  signal, *see* Signal, effective
Electric filters 1
  classical 1–5
  general-purpose 5
  lossless 2, 111
  modern 1–5
  passive 2
  switched-capacitor (SC) 5–12
  see also Filter and Low sensitivity of filters
Electronic filters, *see* Electric filters
Elementary sequence, *see* Sequence, elementary
Encoder 94
Energy
  effective 4
  physical 4, 52
  signal 4
  transmitted 2
Enomoto, T. 177, 256
Esteban, D. 99, 253, 256
Etching effect 23, 25
Even–odd delay element, *see* Delay element, even–odd
Expansion, sampling rate, *see* Up-sampling

FALCON program 116
Fang, S.C. 129, 268
Fast Fourier transformation (FFT) 55
FDM, *see* Frequency-division-multiplexing
Feldtkeller relations 59
Fettweis, A. 1, 4, 5, 14, 32n, 41, 49, 50, 51, 55, 58, 91, 93, 112, 114, 118, 204, 211, 215, 217, 224, 231, 241, 243, 255, 256–7, 261, 266
FFT, *see* Fast Fourier transformation
Filter 1–5
  active-RC 4
  adaptive, *see* Adaptive filter
  analog 2
  anti-aliasing, *see* Anti-aliasing filter
  anti-imaging, *see* Anti-imaging filter
  attenuation 2, 81, 117, 233, 243
  bandpass 242, 244
  bank, *see* Bank, filter
  Cauer, *see* Cauer filter
  centre frequency 126, 127
  classical LC 11
  coefficient 174, 177, 199
  digital 2, 11
    ladder 55
    lattice 55

see also Wave digital filter
discrimination 118
dynamic range 9, 10, 105, 224, 243
frequency range 10
  limits **11**
frequency translated 243–4
group delay 81, 97
highly-selective 221, 241–4
high-pass 59, 93–5, 99
low-pass 59, 64, 85, 93–5, 106, 234–40
lossless 2–3, 50, 111
moving-target-indicator 119
notch, *see* Notch filter
N-path, *see* N-path filter
optimization 233–40
order 118
passive 2
poles 9
postmodulation, *see* Postmodulation filter
premodulation, *see* Premodulation filter
pseudo-N-path, *see* Pseudo-N-path filter
sampled-data 2
SC, *see* Switched-capacitor filters
selectivity 10, 221
tolerance scheme **81**
transversal, *see* Finite impulse response filter
VIS-SC, *see* Voltage inverter switch switched-capacitor circuits
wave digital, *see* Wave digital filter
Wiener, *see* Wiener filter
zeros, *see* Transmission zeros
Finite impulse response (FIR) filter 8, 86, 100, 105, 106, 247
  delay line structure 170–1, 186, 187–91, 196–8
    multi-op-amp 170, 187, 195
    multiphase 187, 190, 195
    reversed 170, 171
  parallel structure 170, **172**, 173, 186, 195
    see also Multi-C FIR structure
  rotator structure 170, **172**, 173, 187, 192, **194**, 195, 197–8
  SC realization 169–202, 203
    basic structures 186–92, 196–7
    composite structures 192–8
    morphological design 186–98
    multirate, *see* Multirate, FIR SC filters
    see also Morphological approach
Finite pseudoenergy signal 53
  see also Pseudoenergy
Fino, M.H. 129, 134, 258
FIR filter, *see* Finite impulse

278 *Index*

response filter
Fischer, G. 169, 177, 190, 258
Flanagan, J.L. 91, 253
Fleischer, P.E. 15, 258
Fliege, N. 12, 100, 107, 157, 258
Forward Euler transformation *206*
Fourier discrete-time signal representation 32n, 35–8
  see also Fourier transformation
Fourier transform 13, 35, 79
  see also Spectrum
Fourier transformation
  discrete (DFT), *see* Discrete Fourier transformation
  discrete-time (DtFT) 35–8
  inverse 35, 37–8, 58
    see also Inverse discrete Fourier transformation
  'strictly' pseudolossless 38n
Fraiture, L. 231, 232, 258
Franca, J.E. 14, 129, 134, 169, 177, 197, 200, 203, 221, 244, 256, 258–9, 263
Franklin, G.F. 43, 267
Franks, L.E. 242, 259
Frequency-division-multiplexing (FDM) 93, 241
Frequency
  response
    precise tuning 119, 127–8
    shifting 119, 126–8
  transformations *206*
    see also Bilinear and Backward, Forward, and Reversed forward Euler transformations
  translation technique 242–5
    see also Filter, frequency translated
Fringe effect 23
Fukasawa, S. 169, 264
Function
  analytic 45
  characteristic, *see* Characteristic function
  conversion, *see* Conversion function
  rational with real coefficients 98n, 99, 116

GaAs technology, *see* Gallium arsenide technology
Galand, C. 99, 253, 256
Gallium arsenide (GaAs) technology 9, 11, 135
Ganesan, A. 15, 258
Gaussian plane 35n, 43, 45
Gazsi, L. 105, 116, 259
Gear, C. 51, 259
Generalized multirate system, *see* System, multirate, general concept
General node 156
  splitting **163**
Gersho, A. 33, 259
Ghaderi, M.B. 14, 204, 259, 263
Gillingham delay element, *see* Delay element, Gillingham

Gillingham, P. 177, 259
Givens rotation 55, 58
Givens, W. 55, 260
Gold, B. 101, 267
Gray, A.H. 55, 260
Gray, P.R. 5, 14, 261, 268
Gregorian, R. 14, 23, 260
Grimbleby, J.B. 129, 260
Group delay, *see* Filter, group delay
Grünigen, D.C. 14, 177, 242, 260
Gyrator 2, 213n

Haigh, D.G. 12, 135, 244, 259, 260, 261, 268
Half-band filters, *see* Complementary filters, half-band
Handkiewicz, A. 204, 260
Harrold, S.J. 135, 260, 261
Hasler, M. 144, 261
Herbst, D. 215, 217, 257, 261, 262
Herley, C. 12, 111, 269
Heteromerous sampling 13, 32–3, 45, 46, 51, 61, 73–4, 119–28
  commensurate 21, 120–1
  second-order 33, 40, 200
Höfflinger, B. 217, 257, 261, 262
Hökenek, E. 135, 204, 261
Hong, Z. 248, 261
Horvat, H. 26, 268
Hostička, B.J. 5, 261
Huang, Q. 248, 261
Hurst, P.J. 119, 126, 251, 261

Hurwitz polynomial 117, 232

IDFT, *see* Inverse discrete Fourier transformation
IIR filter, *see* Infinite impulse response filter
Impedance
  canonic (lattice) 112–3
  normalized input 232
  pulse 230
Impulse response 76
Inductor 3
Infinite impulse response (IIR) filter 86, 102, 105, 106, 247, 248
  direct design 204
  indirect design 204
  voltage–charge modelling 204, 205–11
  wave modelling 204
  SC realization 203–40
Insertion filter, *see* LC ladder, doubly resistively terminated
Instability, numerical 13
Integer
  bands 40–1
  L-bands 41–2
Integrated circuits
  hybrid 3
  low-voltage 4, 11
  low-power 4, 11
  mixed analog-digital 5, 244, 245
  monolithic 4
  see also Active circuits

Integrator 2, 15, 248
Interconnection
   metal 26
   polysilicon 26
Interpolation 12, 62–4, 119, 224, 227–30
   see also Interpolator
Interpolator 14, 20, 104, 170, 203, 221, 225, 235–40, 244
   heteromerous **123**
   IIR SC 200–2
   linear 201–2
   polyphase structure **90**, 110, 200
   scheme **62**
   see also Interpolation
Inverse discrete Fourier transformation (IDFT) 37–8, 58, 104, 105, 110
   two-point **103**, 104, **113**, **115**
Inverter, 15
Ishihara, T. 177, 256
Ishii, J. 117, 266

Jahanbegloo, M. 204, 205, 268
Jain, V.K. 101, 261
Jayant, N.S. 94, 262
Jerri, A.J. 40, 262
Johns, D.A. 248, 262
Johnston, J.D. 98n, 101, 262
Junction
   leakage effect 18
   p-n 18

Kaelin, A. 204, 262
Kim, C. 248, 268

Kleine, U. 51, 204, 217, 226, 262
Klimkiewicz, W. 32, 270
Kohlenberg, A. 40, 262
Kohlenberg sampling theorem 40–1
Kończykowska, A. 129, 134, 252, 262
Korzec, Z. 119, 126, 169, 262
Kotelnikov, V.A. 12, 39, 263

Ladder structure
   LC, *see* LC ladder
   orthonormal 248, **249**
   with unit elements 218
   see also Multirate ladder arrangement
Laker, K.R. 15, 258
Lattice
   impedance, *see* Impedance, canonic (lattice)
   structure 113
      bireciprocal 114, 116
      see also Multirate lattice arrangement
Law, S. 14, 259
LBRF, *see* Linear bounded real filter
LC ladder 1
   doubly resistively terminated 2, 50, 111, 207, 211, 212
   low sensitivity of 3–4, 55, 211, 240
   SC simulation of 205–11
   VIS-SC realization of 235–6
   see also Low sensitivity of

filters
Least-mean-square (LMS) adaptation algorithm 247, 248, **249**
Lee, Y.S. 51, 169, 182, 204, 207, 263
Leickel, T. 91, 257
Leontiev, A.P. 45, 263
Lewis, S.H. 251
Liapunov function 1, 4, 55
    see also Pseudoenergy
Lin, P.M. 129, 263
Linear bounded real filter (LBRF) 55
Linear phase transfer function, see Transfer function, linear phase
LMS, see Least-mean-square adaptation algorithm
Loading resistance 2
    see also Resistor
Looped conditions stability, see Stability under looped conditions
Losses of pseudoenergy, see Pseudoenergy, losses
Losslessness 1
Lowenschuss, O. 119, 267
Low-pass/high-pass transformation 106
Low sensitivity of filters 3–4, 50, 55, 240
Ludwig, M. 14, 260
Lumped element 232

Magnitude distortion 56, 97, 98, 101, 104–5
    see also Subband coding
Mallat, S.G. 111, 263
Maloberti, F. 6, 14, 33, 204, 246, 263, 264, 267
Markel, J.D. 55, 260
Martin, K.W. 16, 23, 169, 182, 260, 263
Martins, R.P. 14, 169, 203, 259, 263
Mason's formula 89n, 141, 142, 155, 168
Mason, S.J. 135, 264
Matsui, K. 169, 264
Matsuura, T. 169, 264
Mavor, J. 51, 204, 226, 264
McClellan, J.H. 106, 264
McConnell 119, 264
McWhirter, J.G. 55, 264
Meerkötter, K. 55, 224, 257
Memory elements 170, 171–82
    see also Delay element, Recharge element and Sample-and-hold circuit
Menthor Graphics Filter Architect software package 116
Menzi, U. 14, 131, 153, 169, 177, 178, 182, 187, 196, 247, 248, 255, 264
Metal interconnection 26
Metal-oxide-semiconductor (MOS) technology 5, 11
    typical data 7
MFSK, see Minimum frequency-shift keying
Minimum frequency-shift key-

ing (MFSK) 169
Mikula, J. 134, 256
Mirror reflection 72
Mitra, S.K. 55, 201, 203, 221, 246, 247, 259, 265, 266, 267, 269
Modified $\mathcal{Z}$ transfer function, *see* Transfer function, modified $\mathcal{Z}$
Modulation
    product, *see* Product modulation
    sequence, *see* Sequence, modulation
    see also Modulator
Modulator
    product, *see* Product modulator
    sigma-delta 15, 246, 248
    see also Modulation
Montecchi, F. 204, 264
Morphological approach 170, 194–6
Morphological box 195
Morris, S. 6, 14, 267
MOS technology, *see* Metal-oxide-semiconductor technology
Moschytz, G.S. 14, 129, 131, 134, 135, 144, 153, 155, 169, 177, 178, 182, 187, 194, 196, 204, 247, 248, 255, 260, 261, 264, 265
Mulawka, J.J. 129, 134, 177, 265
Multi-C FIR structure 186, 192,
193, 195, 197–8, 199
Multiplier circuit 15, 171, 182–6, 248
Multirate
    FIR SC filters 199–202, 246, 247
    frequency translated filters 243–4
    IIR SC arrangements 203–4, 247
    ladder arrangement 118, 204
    lattice arrangement 112–7
    period 13
    systems, *see* System, multirate
Multiresolution
    coefficients 13
    signal representation, *see* Wavelet transformation

Nagaraj delay element, *see* Delay element, Nagaraj
Nagaraj, K. 130, 177, 265
Napieralski, A. 169, 252, 265
Nathan, A. 161, 265
Nathan's rules 159
Neirynck, J.J. 231, 232, 258
Neuvo, Y. 201, 265, 267
Nicholson, W.E. 14, 260
Noble identities 84–5
Node
    general, *see* General node
    sink, *see* Sink node
    source, *see* Source node
Noise shaping technique 245–6
Noll, P. 94, 262

Nonrecoverable aliasing 36, 37, 42, 69, 96, 246
Norator 144, 158
Nossek, J.A. 112, 118, 204, 215, 224, 257, 265
Notch filter 126
Noullet, J.L. 169, 252, 265
N-path filter 242–4
  see also Pseudo-N-path filter
N-stage shift register 243
Nullator 144, 158
Nyquist, H. 12, 39, 265
Nyquist frequency 39, 242, 243, 245

OFR, see Recharge element, open-floating-resistor
Op amp, see Operational amplifier
Operational amplifier 2, 235
  bandwidth 8, 27
  CMOS 27
  DC offset-voltage 132, 133
  differential-input 161, **162**
  finite DC gain 27, 132, 133, 144, **162**, 173
  gallium arsenide 27
  inverting 145, **162**
  monolithic integrated 3, **28–30**
  multiplexing 169, 187
  see also Operational transconductance amplifier
Operational transconductance amplifier (OTA) 27–30
  bias current *30*
  cascode **29**, **30**
  DC gain *30*
  gain-bandwidth product *30*
  layout **28–30**
  Miller **28**
  phase margin *30*
  power consumption *30*
  technical data *30*
  see also Operational amplifier
Operation period 49
Operation rate 21, 49
Oppenheim, A.V. 38, 49, 53, 70, 169, 266
Orchard, H.J. 1, 204, 205, 266, 268
Orthogonal filter, see Power wave digital filter
OTA, see Operational transconductance amplifier
Overlapping of spectra, see Nonrecoverable aliasing
Oversampling technique 245–6
  see also Data converters
Oxide layer 22
Ozaki, H. 117, 266

Pain, B.G. 177, 266
Pandel, J. 215, 217, 243, 261, 266
Parasitic oscillations 55, 58, 60
Paraunitariness 60, 221
  see also Pseudolosslessness
Paraunitary matrix 60
Paraunitary system 55, 60, 99,

284 *Index*

    111, 221
    see also Pseudolossless system
Parks, T.W. 106, 264
Parseval equation 38n, 53, 57, 70, 222, 223
Passband
    ripple 27, 233–4, 235
    sensitivity 2-3, 55, 58, 60, 105, 240
PCM, *see* Pulse-coded modulation
PCTSC, *see* Recharge element, parasitic-compensated-toggle-switched-capacitor
PDM, *see* Pulse density modulation
Perfect reconstruction SBC scheme, *see* Subband coding, perfect reconstruction scheme
Periodically time-varying system, *see* System, periodically time-varying
Periodic $\mathcal{Z}$ transfer function, *see* Transfer function, periodic $\mathcal{Z}$
Petraglia, A. 246, 266
Phase distortion 98
    see also Subband coding
Phoong, S-M. 32, 40, 269
Piekarski, M.S. 55, 267
Piezoelectric effect 2
p-n junction, *see* Junction p-n
Polynomial, Hurwitz, *see* Hurwitz polynomial
Polynomials, relatively prime 116
Polyphase
    components 46, 47, 49, 89, 104–5, 116, 120, 125, 136, 146, 192, 194, 199, 203, 222
    see also Reflectance, canonic
    decomposition of discrete-time signals 33, 46–9, 121, **124**, 222, 245, 246
    see also Signal
    representation of discrete-time systems 85–90, 203, 218, 247
    multistage 192
    nonuniform 121–6
    single-stage 193
    see also System and Transfer function
Polysilicon
    interconnection 26
    layer 22
Postmodulation filter 74, 80, 82, 221, 224, 225
Power
    consumption 6, 9–10
    incident 2, **3**
    reflected 2, **3**
    transmitted **3**
Power complementary
    filter bank, *see* Quadrature-mirror filter bank
    condition 98, 105, 116

Power wave 58
Power wave digital filter (PWDF) 55
Premodulation filter 74, 82, 221, 225
Product
  modulation 12, 46, 61, 67–74
  modulator **68**, 241
Proudler, I.K. 55, 264
PR SBC scheme, *see* Subband coding, perfect reconstruction scheme
Pseudoenergy 52–5
  effective 221, 225
  losses 61, 221–4
    owing to filtering 221–2, 224
    owing to modulation 222–4
  recovery 60, 204, 224–40, 241, 243
  reference, *see* Reference pseudoenergy
  sequence, *see* Sequence, pseudoenergy
  signal, *see* Signal, pseudoenergy
  transmission 221–40
  wave, *see* Wave pseudoenergy
Pseudolosslessness 55–60, 221
  see also Paraunitariness
Pseudolossless system 4n, 57–60, 111, 221
  see also Paraunitary system
Pseudolossless transforms 38n
  see also Parseval equation

Pseudo-N-path filter 242–4
  see also N-path filter
Pseudopassive system 4, 49–60
Pseudopassivity 55–60
Pseudopower 52–7
  reference, *see* Reference pseudopower
  reflected **3**, 234
  sequence average, *see* Sequence, average pseudopower
  signal, *see* Signal, pseudopower
  wave, *see* Wave pseudopower
Pulse-coded modulation (PCM) 93n
  see also Time-division-multiplexing
Pulse density modulation (PDM) 245
Pulse impedance, *see* Impedance, pulse
Pulse width modulation (PWM) 245
PWDF, *see* Power wave digital filter
PWM, *see* Pulse width modulation

Q-factor 10–11, 26, 27, 244
  limits **11**
QMF bank, *see* Quadrature-mirror filter bank
Quadrature-mirror filter (QMF) bank 98–117, 221, 245, 246

Quantization 33–4
  error 34
  levels 33, 34
  noise 33, 245
  resolution 34
  steps 33
Quasi-polynomial 232

Rabaey, J. 129, 269
Rabiner, L.R. 12, 40, 101, 253, 267
Ragazzini, J.R. 43, 267
Rahim, C.F. 252
Rate
  operation, see Operation rate
  sampling, see Sampling rate
Rational function, see Function, rational
Reactance 113
  canonic (lattice) 117, 118, 218, 220
  synthesis 117, 218
Recharge element 170, 173–7, 199
  open-floating-resistor (OFR) **175**, 176, *206*
  parasitic-compensated-toggle-switched-capacitor (PCTSC) **175**, 176–7, 186, 199, *206*
  toggle-switched-capacitor (TSC) **175**, 176–7, 186, 199, *206*
  toggle-switch-inverter (TSI) **175**, 176, 199, *206*

Recharge summer circuit, see Summer circuit, recharge
Reconstruction
  formula 39, 41
  signal, see Signal, reconstruction
Recovery of effective pseudo-energy, see Pseudoenergy, recovery
Recursive least squares (RLS) adaptive algorithm 55
  see also Adaptive filter
Reekie, H.M. 51, 204, 226, 264
Refai, S. 129, 134, 135, 138, 167, 251
Reference circuit 50, 51–2, 111, 204, 205, 225–6, 231
  continuous-time 50
  discrete-time 50–1
  ladder structured 112, 118, 218, 235
  lattice structured 112, 218
  lossless 50, 204, 231
  low sensitivity of 58, 204
  port **50**
  reciprocal 231
  secondary 207
  SSN-type **208**
Reference pseudoenergy 54
  partial 54
  total 54
Reference pseudopower 54
  instantaneous 54
Reflectance
  canonic 113–7, 118, 218, 220
  all-pass 113–7

Index 287

bireciprocal 114
  computation 116–7
Resistance
  equivalent 7
  wave, *see* Characteristic constant
  see also Resistor
Resistor
  monolithic 4
  printed 3
  SC realization of 6–7
  see also Resistance
Reversed forward Euler transformation *206*, 207, 210
Richards, P.L. 117, 267
Riemann surface 35n, 44
RLS, *see* Recursive least squares
r.m.s. value 57
Rotator switch 170
Roundoff noise 55, 58
Roy, R. 119, 267

Saal, R. 235, 267
Said, A.E. 177, 267
Sample-and-hold circuit 170, 173, 200, 242n
Sampling
  bandlimited theory 40–2
    see also Kohlenberg sampling theorem
  classical theorem, *see* Classical sampling theorem
  commensurate 21, 120–1, 135
  heteromerous, *see* Heteromerous sampling
  jitter 246
  Kohlenberg theorem, *see* Kohlenberg sampling theorem
  nonuniform 13, 119–28
    periodic, *see* Heteromerous sampling
  period 32, 44
  'physical' instants 32n
  process 31–33
  rate 32, 44
    alteration 12, 61–7, 74, 169, 224
    compression, *see* Downsampling
    compressor, *see* Downsampler
    expander, *see* Up-sampler
    expansion, *see* Up-sampling
  theorems 38–42
    see also Classical and Kohlenberg sampling theorems
  uniform 32, 51, 52
Sanchez-Sinencio, E. 182, 251
Sandberg, I.W. 242, 259
Santos, S. 14, 197, 200, 259
Saramäki, T. 201, 267
Sauvagerd, U. 91, 257
SBC, *see* Subband coding
SCADAP FIR/IIR adaptive filter module 248–9, **250**
Scattering
  matrix 58–60

transfer functions 58–60, 113
SC circuits, *see* Switched-capacitor circuits
Schafer, R.W. 38, 49, 53, 70, 169, 266
Schumacher, K. 217, 261
Scott, N.L. 40, 269
Secondary prototype, *see* Reference circuit, secondary
Sedra, A.S. 248, 262
Sensitivity 1
   passband, *see* Passband sensitivity
   stopband, *see* Stopband sensitivity
   see also Low sensitivity of filters
Sequence
   average pseudopower 70
   elementary 70, 71, 224, 241
      sinusoidal *72*
   essentially equivalent 72–3
   essentially different 72–3
   modulation (carrier) 13, 68, 69, 70, 73, 222, 225
      sinusoidal 71
   pseudoenergy 69–70
Settling time 28
SFG, *see* Signal-flow graph
Shannon, C.E. 12, 39, 267
Shen, Z-G. 134, 270
Shifting the frequency response, *see* Frequency response, shifting
Shynk, J.J. 247, 248, 267
SI circuits, *see* Switched-current circuits
Sigg, R. 14, 260
Signal
   acoustic 91
   bandlimited 40
   causal 53
   carrier 13, 67, 69, 73, 74
      sinusoidal 70–2
   complex modulated 48–9
   continuous-time, *see* Continuous-time, signal
      spectrum 39
   discrete-time, *see* Discrete-time, signal
   polyphase representation, *see* Polyphase, decomposition of discrete-time signals
      spectrum **37**
   Doppler, *see* Doppler signal
   effective 65, 74, 221
   finite pseudoenergy, *see* Finite pseudoenergy signal
   integer-band 40–1
   integer L-band 41–2, 64, 67, 69, 75, 222, 223, 243
   lowband 39
   non-bandlimited 40
   perfect reconstruction, *see* Subband coding, perfect reconstruction scheme
   polyphase
     components, *see* Polyphase, components
     decomposition, *see* Polyphase, decomposition of

Index   289

discrete-time signals
processing
  adaptive 13
  see also Adaptive filter
  statistical 13
pseudoenergy 53, 222, 224
  partial 53
  total 53
pseudopower 53
  instantaneous 53
radar 32, 119
real-valued 53
reconstruction 32, 68, 92, 94, 119
sampling 31–33, 38–42
sinusoidal 71
sonar 32
splitting into subbands, *see* Subband coding
step-wise (sampled-and-held) 31, 200, 201, 221, 242n, 244
Signal-flow graph (SFG) 49, 51
  analysis of SC networks 129, 134–68
  concepts 135–9
    Bedrosian and Refai approach **137**, 138–43
    Coates approach **137**, 138–43
    Mason approach **137**, 138–43
  connection 141
    gain 141
  delay-free loop 49n, 50, 51
  digital 50

directed branch 49n, 140
directed loop 49n, 141
  composite 141
  disjoint 141
  gain 141
  joint 141
discrete-time 50
node 140
path 140
  gain 140
realizable 50
self-loop 138, 141
  gain 138, 139
wave port **50**
Signal-to-noise ratio (SNR) 8, 10, 34
Singhal, K. 129, 269
Sink node 144, 146, 150, 153, 158, 174
Sink–source node pairing 153, 161
Smith, M.J.T. 106, 267
Snelgrove, M. 248, 262
SNR, *see* Signal-to-noise ratio
Soin, R.S. 6, 14, 267
Somayazulu, V.S. 247, 267
Song, H. 248, 268
Source node 144, 146, 150, 153, 158
  dependent 144
  independent 144
Source-sink-node-network (SSN-network) transformation 156–63, 207
Source-sink-node-type (SSN-type)

active-RC network 144n, 205
SC network 144–56
Spectrum
   bandwidth 12
   continuous-time signal 39
   conventional 36
   discrete-time signal **37**
   magnitude 36
   modified 36
   see also Fourier transform
SSN-network transformation,
      see Source-sink-node-network transformation
SSN-type SC network, *see* Source-sink-node-type SC network
Stability 55, 105, 224
   analysis 55
   margin 55, 224
   structural (robust) 4, 50, 204, 211
   under looped conditions 55, 224
      see also Communication link
   see also Liapunov function
Static system 57n, 58
   see also Adaptor
Steady-state 52, 75, 119, 229
Steiger-Garção, A. 129, 134, 258
Stopband
   attenuation 225, 233, 235
   sensitivity 3, 218
Stub 24–25
Subband coding (SBC) 56, 91–111, 221

aliasing-free scheme 96–7, 105
biorthogonal scheme 107
four-channel scheme **107**
magnitude preserving scheme 97, 105
   see also Magnitude distortion
multichannel scheme 108–11
perfect reconstruction (PR) scheme 56, 98, 101–7, 110
phase preserving scheme 98
tree-structured scheme 107–10
two-channel scheme 94–107, 218, 221
Substrate (Si) 22
Summer circuit 15, 170, 171, 182–6, 248
   Lee–Martin 182–5, 187, 192, 198
   recharge 182, **184**, 185–6, 187, 198
   see also Multiplier circuit
Surface acoustic wave effect 2
Suyama, K. 129, 268
Switch 16–18, 20
   circuit **17**
   layout **17**
   off-resistance 17
   on-resistance 16, 17, **18**
Switched-capacitor (SC) circuits 5–12, 111
   basic elements 15–30
   insensitive to capacitor mismatch 131, 132, 178

multiphase 14
  analysis 129–68
multirate 14, 111, 199–202, 246
offset-compensated 131, 178, 185, 248
parasitic compensated 15, 144n, **175**, 176–7, 186
parasitic insensitive 15, 131–2, 135, 144, 151, 178, 240
parasitic sensitive 15, 131, 205
stray-insensitive, *see* parasitic insensitive
see also Switched-capacitor (SC) filters
Switched-capacitor (SC) filters 5–12
  FIR, *see* Finite impulse response filter, SC realization
  frequency translated 243–4
  high order 26
  high Q 26
  high selectivity 26
  IIR, *see* Infinite impulse response filter, SC realization
Switched-current (SI) circuits 8
Synthesis filter bank 14, 60, **91**, 92, **94**, 96–115, 246, **247**
System
  adaptive, *see* Adaptive filter
  continuous-time 50, 52, 244
  discrete-time 31–60, 111

externally pseudolossless 57
externally pseudopassive 57
full-synchronic 49
half-synchronic 49
internally pseudolossless 58, 105
internally pseudopassive 58
paraunitary, *see* Paraunitary system
  see also Pseudolossless system
pseudoenergy neutral 57
pseudolossless, *see* Pseudolossless system
  see also Paraunitary system
pseudopassive 4, 49–60
wave, *see* Wave discrete-time system
dynamic, *see* Dynamic system
linear 76, 135
low-dynamic range 9, 11, 224
multirate 5, 12–14, 55, 61–118
  analysis 75–82
  applications 13
  construction of frequency characteristics 79–82
  equivalences for building-block connections 82–5
  see also Noble identities
  general concept 61, 73–5,

221
  low-pass 81–2, 234–40
  polyphase representation, see Polyphase, representation of discrete-time systems
  scheme of **75**
  periodically time-varying 75–9, 82
  shift-invariant (stationary) 51, 85
  single-rate 51
  static, see Static system

Taylor, R.C 26, 268
TDM, see Time-division-multiplexing
Temes, G.C. 14, 23, 204, 205, 259, 260, 263, 265, 268
Time constant 7
Time-division-multiplexing (TDM) 93, 241
  see also Pulse-coded modulation
Time-varying $\mathcal{Z}$ transfer function, see Transfer function, time-varying $\mathcal{Z}$
Timing circuit, see Clock
Topological formulae 135, 139–43
  see also Bedrosian–Refai formula, Coates–Desoer formula and Mason's formula
Toumazou, C. 135, 260, 268
Tounsi, L. 119, 126, 262

Transient analysis 119
Transmittance, see Transfer function
Transfer function 5n, 10, 58
  all-pass, see All-pass transfer function
  conventional $\mathcal{Z}$ 85–6
  current 230
  distortion, see Distortion transfer function
  finite impulse response (FIR) 86, 170
  infinite impulse response (IIR) 86–9, 205
  inherently stable 169, 170
  linear phase 97, 100, 107, 169
  modified $\mathcal{Z}$ 86
  nonrecursive, see finite impulse response
  normalized 117
  open-circuit voltage 230
  periodic 77
  polyphase realization 86–9, 102, 104, 203
    type 1 structure **88**, 89, 90
    type 2 structure **88**, 89, 90
  recursive, see infinite impulse response
  scattering, see Scattering transfer functions
  time-varying $\mathcal{Z}$ 76
  zero-phase 101
Transfer functions of SFG branches *151*
Transistor (MOS)

n-type 17, 20
p-type 17, 20
Transition band 27, 102
Transmission gate, *see* Switch
Transmission zeros 1, 247
Transmittance, *see* Transfer function
Transmultiplexer 13, 41, 93, 241, 242
Trapezoidal rule 51, 52
Treichler, J.R. 247, 268
TSC, *see* Recharge element, toggle-switched-capacitor
TSI, *see* Recharge element, toggle-switched-inverter
Tsividis, Y.P. 32n, 268
Tukey, J.W. 55, 252
Tuszyński, M. 32, 45, 270

Uehara, C.T. 14, 268
Unbehauen, R. 129, 134, 268
Unit element 213n, 218, 220, 231, 232
   see also Ladder structure
Unit circle 98
Up-sampler 62, 82, 90, 95, 96
   simple equivalences **83**
Up-sampling 62, 73, 224

Vaidyanathan, P.P. 12, 14, 32, 40, 55, 100, 105, 109, 111, 119, 268, 269
Valid half-band filters, *see* Complementary filters, valid half-band
Vance, I.A.W. 135, 261
Vandewalle, J. 129, 269

Vaughan-Pope, D.A. 55, 252
Vaughan, R.G. 40, 269
Very large scale of integration (VLSI) 5
Vetterli, M. 12, 93, 98, 109, 111, 269
Virtual ground **146**
VIS, *see* Voltage inverter switch
VIS-SC circuit, *see* Voltage inverter switch switched-capacitor circuit
Vittoz, E.A. 9, 269
Vlach, J. 129, 269
Vlach, M. 129, 269
VLSI, *see* Very large scale of integration
Vollenweider, W. 242, 260
Voltage inverter switch (VIS) 4, 211, **215**, 228
   floating 215, **217**
   grounded 215, **216**
   inverse recharging type 215, **216**, 220, 227–8
   parasitic insensitive 215, 240
   voltage registration type 215, **216**
Voltage inverter switch switched-capacitor (VIS-SC) circuits 4, 32n, 51, 112, 204, 211–40
   ladder structured 217, 218, 234–40
   lattice structured 217, 218–21
   multirate 227–8, 231
   with pseudoenergy recovery

224–40, 243
Voltage wave digital filter (VWDF) 55
VWDF, see Voltage wave digital filter

Walter, G.C. 40, 269
Wave
  adaptor, see Adaptor
  incident 50, 51n, 213
  power, see Power wave
  reflected 50, 51n, 213, 225, 228
  resistance, see Characteristic constant
Wave digital filter (WDF) 4, 49–51, 55–60, 224–6
  low sensitivity of 50
  see also Wave discrete-time system
Wave discrete-time system 49–60
  multirate 111–8, 224
Wavelet
  analysis 13
  transformation 13, 111
Wave pseudoenergy 54
  partial 54
  total 54
Wave pseudopower 54
  average absorbed 57
  instantaneous 54, 56
Wave-SC circuit 51, 204, 226
WDF, see Wave digital filter
Webber, S.A. 91, 253
White, D.R. 40, 269

Whittaker, E.T. 12, 39, 270
Whittaker, J.M. 12, 39, 270
Wiener filter 248
Wojtkiewicz, A. 32, 45, 270
Wooley, B. 246, 252
Wupper, H. 243, 257

Yasumoto, M.A. 177, 256

Zadeh, L. 75, 270
Zarzycki, J. 55, 267
Zayed, A.I. 40, 270
Zbinden, P. 29, 169, 170, 182, 247, 248, 250, 255, 264, 270
Zero-phase function, see Transfer function, zero-phase
Zhou, X-F. 134, 270
$\mathcal{Z}$ transform 42–6, 48, 58, 136
  conventional 48, 49, 63, 67, 69
  modified 48, 49, 63, 67, 68, 95
  analytic part 44, 45
  delay part 44
  see also $\mathcal{Z}$ transformation
$\mathcal{Z}$ transformation 32n, 42–6
  advanced 43
  conventional (classical) 42, 43
  delayed 43
  modified 42–6
  see also $\mathcal{Z}$ transform
Zwicky, F. 194, 270